Oxford Texts in Applied and Engineering Mathematics

OXFORD TEXTS IN APPLIED AND ENGINEERING MATHEMATICS

Titles marked with an asterisk (*) appeared in the Oxford Applied Mathematics and Computing Science Series, which has been folded into, and is continued by, the current series.

Introduction to Monte-Carlo Methods for Transport and Diffusion equations

B. LAPEYRE

CERMICS, École Nationale des Ponts et Chaussées

É. PARDOUX

Université de Provence

R. SENTIS

Commissariat à l'Énergie Atomique

Translated by Alan Craig and Fionn Craig

OXFORD
UNIVERSITY PRESS

*This book has been printed digitally and produced in a standard specification
in order to ensure its continuing availability*

OXFORD
UNIVERSITY PRESS

Great Clarendon Street, Oxford OX2 6DP

Oxford University Press is a department of the University of Oxford.
It furthers the University's objective of excellence in research, scholarship,
and education by publishing worldwide in

Oxford New York

Auckland Cape Town Dar es Salaam Hong Kong Karachi
Kuala Lumpur Madrid Melbourne Mexico City Nairobi
New Delhi Shanghai Taipei Toronto
With offices in
Argentina Austria Brazil Chile Czech Republic France Greece
Guatemala Hungary Italy Japan South Korea Poland Portugal
Singapore Switzerland Thailand Turkey Ukraine Vietnam

Oxford is a registered trade mark of Oxford University Press
in the UK and in certain other countries

Published in the United States
by Oxford University Press Inc., New York

Oxford is a registered trade mark of Oxford University Press
in the UK and in certain other countries

Published in the United States
by Oxford University Press Inc., New York

Translation from the French language edition:
Méthodes de Monte-Carlo pour les equations de transport et de diffusion
by Bernard Lapeyre, Étienne Pardoux and Rémi Sentis

Copyright © 1998 Springer-Verlag Berlin-Heidelberg

Springer-Verlag is a company in the Bertelsmann Springer publishing group

English edition © Oxford University Press 2003

ISBN 0-19-852593-1

PREFACE

The generic term 'Monte-Carlo method' is used for all numerical methods using sampling of random numbers. These are very useful in numerous fields, in particular in nuclear physics, statistical mechanics, and statistics. Moreover, there are variants in signal processing—called stochastic approximation algorithms, and in optimization—the well known 'simulated annealing' method. In this book, with the exception of the first chapter which concerns the calculation of simple numerical quantities (such as integrals), we shall study these methods only for the numerical solution of particular partial differential equations. In this framework, they are linked to random particle methods.

Although they are becoming popular because, in particular, of the increase in computing power and generalization to the vector and parallel computers, Monte-Carlo methods often have a bad reputation. They are thought to converge slowly and to be not very reliable. Moreover, their mathematical justification is not always clear, and they are often used without having a proof of their convergence. Some of these criticisms have foundation, and the rate of convergence makes these methods useful primarily in situations (and they are numerous) where we do not have another effective numerical algorithm. On the other hand, if Monte-Carlo methods do not always converge (and can demonstrate 'false convergence') *it is always possible*, as we shall see in chapter 1 on a simple example, using an inexpensive additional calculation, *to control the reliability of the result*. This point is absolutely essential and should be well known to all users of Monte-Carlo methods.

As will be clarified in chapter 1, the convergence of Monte-Carlo methods is based on the law of large numbers and interpretation of the quantity to be evaluated as an expected value. To numerically solve partial differential equations by Monte-Carlo (and to be ensured of the convergence of the result), it is therefore appropriate to give a probabilistic representation of the solution of these equations. This aspect of the theory is mostly approached here within the framework of linear equations. We shall therefore seek systematically to interpret transport or diffusion type equations as Fokker–Planck (or Kolmogorov) equations associated with Markov Processes (therefore certain probabilistic formulas for 'Fokker–Planck ' equations could be new). On the other hand, we shall not look further into the theoretical aspects related to the probabilistic representation of nonlinear partial differential equations, in particular, the Boltzmann equation (that would have led to mathematical developments which exceed the objectives of this work). For this last equation, we give only an outline of an approximation result for the solution by expected value of a particular functional, but the systematic study of the convergence of the method is not attempted.

Our text is organized in the following way. Chapter 1 presents the Monte-Carlo method for the calculation of integrals, and studies their properties and limitations. Several basic ideas—essential in the use of the Monte-Carlo methods for the numerical solution of partial differential equations—are introduced here, in particular, the speed of convergence, the control of variance, and the methods for variance reduction. We highly recommend that the reader studies this chapter carefully, even if it appears elementary. Chapter 2 presents the probabilistic interpretation of the transport equations which occur, in particular, in particle physics. Chapter 3 details the principles and the implementation of the Monte-Carlo method for these transport equations. In Chapter 4, we describe the method for the nonlinear Boltzmann transport equation. Chapter 5 presents the links between second-order partial differential diffusion equations and Brownian motion and diffusion processes, then it describes the principle of Monte-Carlo methods for this type of equation.

We have endeavoured, for each type of problem, to emphasize clearly the limits of the method and additionally to describe the specific techniques used in practice (without however going too deeply into the details of the algorithms).

This work is aimed at mathematicians, engineers, and physicists who have a good grounding in analysis (the principal ideas of probability will be introduced as they are used). We hope that it will enable them to understand better the basis of Monte-Carlo methods, as well as the guidelines to be observed for their use.

This work is an amended version of the notes from a course presented by the authors at the 25th Congress of Numerical Analysis, 22 and 23 May 1993. We thank Jean-Marie Crolet, of the Laboratoire de Calcul Scientifique de Besançon, the organizer of the Congress, who invited us to give this course, as well as the listeners who had the courage to follow us on a subject which was in general very new to them. We would also like to thank J. M. Depinay who provided us with a numerical example, and finally A. Cossu who contributed to the typing of this document.

B. L.
E. P.
R. S.

CONTENTS

1

Monte-Carlo methods and integration

The introduction of the Monte-Carlo method can be credited to Count Buffon who, in 1777, described a celebrated method of calculating π based on performing repeated experiments. However, the true origin of the Monte-Carlo method lies with the introduction of the first computers and their use in the secret projects of the United States Department of Defense during the years 1940–5 with the aim of constructing the first atomic bombs. One of the first articles on this subject was published in 1949 (Metropolis and Ulam, 1949). Precursors of these methods include Ulam's method, Von Neumann's method, and the method of Metropolis.

To introduce the Monte-Carlo method, we consider the problem of numerical integration. We know that there are many methods for numerical approximation of the integral:

$$\int_{[0,1]} f(x)\,dx$$

by formulae of the form $\sum_{i=0}^{n} w_i f(x_i)$, where w_i are positive numbers summing to 1 and x_i are points in the interval $[0,1]$. For example, when $w_0 = w_n = 1/(2n)$, $w_i = 1/n$ otherwise, and $x_i = i/n$ we have the trapezoidal method. Many other methods also exist, such as Gaussian integration or Simpson's rule. A Monte-Carlo method has the same form: we choose $w_i = 1/n$ and we take the x_i 'at random' (but not completely arbitrarily, the points must be uniformly distributed over $[0,1]$). This method converges with order $K/\sqrt{n}$. Obviously, this rate of convergence may seem poor if we compare it with other methods in one dimension. But all of these numerical methods break down when the dimension is increased (typically, we have to have n^d points, where d is the dimension, to have a constant error). The great advantage of the Monte-Carlo method is that it is insensitive to the dimension.

1.1 Revision of probability theory

A *random variable* is a function defined over a set Ω which takes values on another set E. We denote by ω a general element of Ω and by X (or another capital letter!) a random variable:

$$X : \Omega \to E.$$

We further assume that this mapping is measurable: Ω is equipped with a sigma algebra $\mathcal{A}$, E with a sigma algebra $\mathcal{E}$, and the mapping X from Ω into E is measurable in the sense that $\{X \in F\} \in \mathcal{A}$ for every $F \in \mathcal{E}$.

In every case that we study, E will be equal to $\mathbf{R}^d$ and $\mathcal{E}$ to the Borel sigma algebra of $\mathbf{R}^d$.

It remains to consider the different realizations ω of Ω. It is this, with the help of a positive measure on $(\Omega, \mathcal{A})$ with total mass 1 usually denoted $\mathbf{P}$, that we call the *probability*.

When X takes values in $\mathbf{R}$ (or more generally in $\mathbf{R}^d$), this measure $\mathbf{P}$ allows us to calculate the *expected value* of X, denoted traditionally by $\mathbf{E}(X)$:

$$\mathbf{E}(X) = \int_\Omega X(\omega) \, d\mathbf{P}(\omega).$$

The expected value is only defined if $E(|X|) = \int_\Omega |X(\omega)| \, d\mathbf{P}(\omega) < +\infty$.

The *distribution* of the random variable X is the image measure of $\mathbf{P}$ under the mapping X. It is a measure over E which we denote by μ_X. The distribution of X under $\mathbf{P}$, μ_X, is characterized by the following property, for every mapping f from E into $\mathbf{R}$ which is measurable and positive (or bounded):

$$\mathbf{E}(f(X)) = \int_E f(x) \, d\mu_X(x).$$

We say that the two random variables X_1 and X_2 are independent if for all positive measurable functions f_1 and f_2 we have:

$$\mathbf{E}(f_1(X_1)f_2(X_2)) = \mathbf{E}(f_1(X_1)) \, \mathbf{E}(f_2(X_2)).$$

Similarly, n random variables $X_1, \ldots, X_n$ are independent if for all measurable positive functions $f_1, \ldots, f_n$ we have:

$$\mathbf{E}(f_1(X_1) \ldots f_n(X_n)) = \mathbf{E}(f_1(X_1)) \ldots \mathbf{E}(f_n(X_n)).$$

Finally, a sequence of random variables $(X_1, \ldots, X_n, \ldots)$ is a sequence of independent random variables if every finite subsequence is independent. For a more complete introduction to probability theory, we refer, for example, to Bouleau (1986) or Breiman (1968).

1.2   Description of the Monte-Carlo method

To use the Monte-Carlo method, we must first put the quantity which we wish to calculate in the form of an expected value. This is often simple, as in the case of integration, but it can be more complicated, for example, when we want to solve a parabolic or elliptic equation, or a linear system. We shall consider this stage at some length later.

At the end of this stage, we have to calculate a quantity of the form $\mathbf{E}(X)$, where X is a random variable. In order to calculate $\mathbf{E}(X)$, we need to know how to *simulate* a random variable under the distribution of X. Mathematically, this means that we assume that we have a sequence of independent random variables $(X_i, i \geq 1)$ all following the distribution of X.

Computationally, we reduce the simulation of an arbitrary distribution to that of a sequence of independent random variables following a uniform distribution on the interval $[0, 1]$ (we shall see in Section 1.6 some ideas about how to simulate the usual distributions). This type of random sequence is often provided in programming languages (rand in C, g05caf and others in NAG, etc.). It only remains to approximate $\mathbf{E}(X)$ by

$$\mathbf{E}(X) \approx \frac{1}{N}(X_1 + \cdots + X_N).$$

We shall give an example of the Monte-Carlo method used to calculate an integral, giving details of the two stages above: transformation into the form of an expectation and simulation of the random variable.

Assume that we want to calculate an integral of the form:

$$I = \int_{[0,1]^d} f(u_1, \ldots, u_d) \, du_1 \ldots du_d.$$

We set $X = f(U_1, \ldots, U_d)$, where $U_1, \ldots, U_d$ are independent random variables distributed uniformly on the interval $[0, 1]$. We have:

$$\mathbf{E}(X) = \mathbf{E}\left(f(U_1, \ldots, U_d)\right) = \int_{[0,1]^d} f(u_1, \ldots, u_d) \, du_1 \ldots du_d,$$

by the definition of the distribution of an n-tuplet, $(U_1, \ldots, U_d)$. We have completed the first stage (transformation into an expected value).

For the simulation, assume that $(U_i, i \geq 1)$ is a sequence of independent random variables uniformly distributed over $[0, 1]$ (obtained by successive evaluation of a **random** function) and set $X_1 = f(U_1, \ldots, U_d)$, $X_2 = f(U_{d+1}, \ldots, U_{2d})$, etc. Then, the sequence $(X_i, i \geq 1)$ is a sequence of independent random variables under the distribution X and a good approximation of I is given by

$$\frac{1}{N}(X_1 + \cdots + X_N).$$

We remark that this method is very easy to program. It is also notable that this method does not depend on the regularity of f, which can be simply measurable.

Often we want to evaluate an integral in $\mathbf{R}^d$, of the form:

$$I = \int_{\mathbf{R}^d} g(x)f(x) \, dx = \int_{\mathbf{R}^d} g(x_1, \ldots, x_d)f(x_1, \ldots, x_d) \, dx_1 \ldots dx_d,$$

with $f(x)$ positive and $\int f(x)\,dx = 1$. Then, I is written in the form $\mathbf{E}(g(X))$ if X is a random variable with values in $\mathbf{R}^d$ with distribution $f(x)\,dx$. We can therefore approximate I by

$$I \approx \frac{1}{N}\sum_{i=1}^{N} g(X_i),$$

if $(X_i,\ i \geq 0)$ is sampled from the distribution $f(x)\,dx$. The problem is then to simulate a random variable under this distribution. For the usual distributions we know how to solve this problem (some currently used methods are given in Section 1.6).

We shall now have to answer two questions:

- How and when does this method converge?
- What can we say about the precision of the approximation?

1.3   Limits and convergence

1.3.1   *Convergence theorems*

The answers to these two questions are given by two of the most important theorems in the calculus of probabilities, the strong law of large numbers, which allows us to justify the convergence of the method, and the central limit theorem, which gives the rate of convergence.

The strong law of large numbers and the Monte-Carlo method

Theorem 1.3.1 *Let $(X_i,\ i \geq 1)$ be a sequence of independent random variables following the same distribution as a random variable X. We assume that $\mathbf{E}(|X|) < +\infty$. Then, for almost every ω (this means that there exists $N \subset \Omega$, with $\mathbf{P}(N) = 0$ and that if $\omega \notin N$):*

$$\mathbf{E}(X) = \lim_{n \to +\infty} \frac{1}{n}(X_1(\omega) + \cdots + X_n(\omega)).$$

This theorem imposes a theoretical limit on Monte-Carlo methods: we can only use it for integrable random variables (which is not really surprising).

The central limit theorem and the Monte-Carlo method For the method to be useful, there must be a way of evaluating the error:

$$\epsilon_n = \mathbf{E}(X) - \frac{1}{n}(X_1 + \cdots + X_n).$$

The central limit theorem gives a quantity that is asymptotically equal to the error ϵ_n but which is random in nature. It says that the distribution ϵ_n resembles a centred Gaussian distribution.

Theorem 1.3.2 *Let $(X_i,\ i \geq 1)$ be a sequence of independent random variables with the same distribution as a random variable X. We assume that $\mathbf{E}(X^2) < +\infty$. We denote by σ^2 the variance of X:*

$$\sigma^2 = \mathbf{E}(X^2) - \mathbf{E}(X)^2 = \mathbf{E}\left((X - \mathbf{E}(X))^2\right),$$

then

$$\frac{\sqrt{n}}{\sigma}\epsilon_n \text{ converges in probability to } G,$$

G being a random variable with a reduced centred Gaussian distribution G.

This means that G is a random variable with distribution $e^{-x^2/2}(dx/\sqrt{2\pi})$ and that if f is a bounded continuous function, $\mathbf{E}(f((\sqrt{n}/\sigma)\epsilon_n))$ converges to $\mathbf{E}(f(G))$.

Remark We can deduce from Theorem 1.3.2 that for all $c_1 < c_2$:

$$\lim_{n \to +\infty} \mathbf{P}\left(\frac{\sigma}{\sqrt{n}}c_1 \leq \epsilon_n \leq \frac{\sigma}{\sqrt{n}}c_2\right) = \int_{c_1}^{c_2} e^{-x^2/2}\frac{dx}{\sqrt{2\pi}}.$$

In practical applications, we 'forget the passage to the limit' and we replace ϵ_n by a centred Gaussian distribution with variance σ^2/n.

Remark We note that the central limit theorem never allows us to bound the error, since the support of the Gaussian is equal to the whole of $\mathbf{R}$. We often describe the error of the Monte-Carlo method by giving the standard deviation of ϵ_n, that is, $\sigma/\sqrt{n}$, or by giving a 95% confidence interval for the result. This means that the result is found with 95% chance in the given interval (and with 5% chance of being outside). Obviously, the value 95% may be replaced by another value close to 1. As

$$\mathbf{P}\left(|G| \leq 1.96\right) \approx 0.95,$$

we are led to a confidence interval of the kind:

$$\left[m - 1.96\frac{\sigma}{\sqrt{n}}, m + 1.96\frac{\sigma}{\sqrt{n}}\right].$$

We must also note that the rate of convergence of the error is $1/\sqrt{n}$, which is not terribly good. However, there exist cases where a slow method is unfortunately the best possible (for multiple integrals in more than 100 dimensions, parabolic equations in 50 dimensions, etc.). It is also remarkable that the rate of convergence of the method, for calculating integrals, does not depend on the regularity of f.

1.3.2 *Estimate of the variance of a calculation*

The result above shows that it is important to know the order of the size of the variance σ of the random variable that we have calculated with the help of the Monte-Carlo method, since this gives an idea of the error in the calculation. It is easy to estimate this variance. We shall show that there are many techniques to reduce the variance (see Section 1.4) and in certain cases the application of one or other of these techniques is indispensable if we want to obtain a reliable result without having to make a prohibitive number of evaluations.

Using the notation of the preceding paragraph (X is a random variable with values in $\mathbf{R}^d$ with distribution $f(x) \, dx$), we want to evaluate an integral of the type:

$$I = \int_{\mathbf{R}^d} x f(x) \, dx = \mathbf{E}(X).$$

Let X_i be independent realizations under the distribution of X, we can therefore approximate I by $\bar{I}_N$ by

$$\bar{I}_N = \frac{1}{N} \sum_{i=1}^{N} X_i,$$

when $N \to \infty$. It is well known that we can also directly obtain an estimate of the variance of X using the formula:

$$\mathbf{V} = \frac{1}{N-1} \sum_{i=1}^{N} \left(X_i - \bar{I}_N \right)^2,$$

$\mathbf{V}$ is often called the empirical variance of the sample. We can therefore obtain a 95% confidence interval by setting $\bar{\sigma} = \sqrt{\mathbf{V}}$ and by replacing σ by $\bar{\sigma}$ in the confidence interval given by the central limit theorem. We therefore obtain a confidence interval for I:

$$\left[\bar{I}_N - \frac{2\bar{\sigma}}{N}, \bar{I}_N + \frac{2\bar{\sigma}}{N} \right].$$

We therefore see that, with almost no extra calculation (just by evaluating $\bar{\sigma}$ on the sample already taken), we could give a dependable estimate of the approximation error of I by $\bar{I}_N$. It is one of the greatest advantages of the Monte-Carlo method to give a realistic estimate of the error at minimum cost.

1.3.3 *Some significant examples*

We shall give some examples of the application of the central limit theorem. We shall see that this result imposes some practical limit on the Monte-Carlo method.

An easy case. Let f be a measurable function defined over $[0, 1]$ and assume that we want to calculate $p = \int_{\{f(x) \geq \lambda\}} dx$ for a given constant

λ. Introduce the random variable $X = \mathbf{1}_{\{f(U) \geq \lambda\}}$ (where U is a random variable uniformly distributed on $[0, 1]$).

Then, $p = \mathbf{E}(X)$ and $\sigma^2 = \text{Var}(X) = p(1 - p)$. Therefore, taking n independent samples following the distribution of X, $X_1, \ldots, X_n$ we have:

$$p_n = \frac{X_1 + \cdots + X_n}{n} \approx p + \frac{\sigma}{\sqrt{n}} G.$$

As $p(1 - p) \leq 1/4$, we see that the standard deviation of the error $\sigma/\sqrt{n}$ is bounded by 0.01, which means that we must take n to be of the order of 2500. If we choose $n = 2500$, the 95% confidence interval for p is then, using the central limit theorem, $[p_n - 1.96 \times 0.01, \; p_n + 1.96 \times 0.01]$. If the value of p to be estimated is of order 0.50 this leads to an acceptable error.

On the other hand, when the value of p to be estimated is very small the number of samples above is obviously insufficient to evaluate the order of magnitude by simulation. We must (and this is intuitively obvious) take a number of samples clearly greater than $1/p$.

A difficult case. Imagine that we want to calculate $\mathbf{E}\left(\exp(\beta G)\right)$, where G is a reduced centred Gaussian random variable. It is easy to verify that

$$E = \mathbf{E}\left(e^{\beta G}\right) = e^{\beta^2/2}.$$

If we apply a Monte-Carlo method, we set $X = e^{\beta G}$. The variance of X is $\sigma^2 = e^{2\beta^2} - e^{\beta^2}$. At the end of n samples following the distribution of X, $X_1, \ldots, X_n$ we have:

$$E_n = \frac{X_1 + \cdots + X_n}{n} \approx E + \frac{\sigma}{\sqrt{n}} G',$$

the random variable G' following a reduced centred Gaussian distribution. The average relative error is of the order of $\sigma/E\sqrt{n} = \sqrt{(e^{\beta^2} - 1)/n}$. If we choose an order of magnitude ϵ for the error to obey, we see that this means that we choose $n \approx (e^{\beta^2} - 1))/\epsilon^2$. If $\epsilon = 1$ and $\beta = 5$, we get $n = 7 \times 10^{10}$, which is (too!) large. Given below, for example, is the result given by a program that estimates this value in the case $\beta = 5$.

exact value	:	268 337
100 000 samples, estimated value	:	854 267
estimated 95 % confidence interval	:	[-467 647,2 176 181] !

We note that this approximation is very misleading. But, importantly, the calculated confidence interval contains the exact value. This is the reassuring aspect of the Monte-Carlo method: the approximation may be mediocre but we are well aware of it!

This example displays a practical limit to the Monte-Carlo method when we use random variables with large variance.

A more concrete example. In financial applications, we have to calculate quantities of the type:

$$C = \mathbf{E}\left(\left(e^{\beta G} - K\right)_+\right),\qquad(1.1)$$

G being a reduced centred Gaussian and $x_+ = \max(0, x)$. These quantities represent the price of an option to buy, commonly called a 'call'. Obviously, in this case we can find an explicit formula (this is the celebrated Black–Scholes formula, Lamberton and Lapeyre, 1991):

$$\mathbf{E}\left(\left(e^{\beta G} - K\right)_+\right) = e^{\beta^2/2} N\left(\beta - \frac{\log(K)}{\beta}\right) - KN\left(-\frac{\log(K)}{\beta}\right),$$

with $N(x) = \int_{-\infty}^{x} e^{-u^2/2}\, du$. But we are going to assume that we want to calculate these quantities by simulation.

The practical values used for β are of the order of unity. In the numerical experiments we have set $\beta = 1.0$ as $K = 1.0$. The results of the simulation and also an estimate of the error for various numbers of trials are given below.

```
exact value : 6.720
N = 100,    95% confidence interval : [0.08,11.39]
            estimated value : 5.74
N = 1000,   95% confidence interval: [4.20,10.01]
            estimated value : 7.1
N = 10000,  95% confidence interval: [6.13,8.43]
            estimated value : 7.28
N = 100000, 95% confidence interval: [6.59,7.69]
            estimated value : 7.14
```

Let us now compare the results with those we have obtained when we have evaluated an option to sell, or 'put', that is,

$$P = \mathbf{E}\left(\left(K - e^{\beta G}\right)_+\right).\qquad(1.2)$$

The explicit formula gives:

$$KN\left(\frac{\log(K)}{\beta}\right) - e^{\beta^2/2} N\left(\frac{\log(K)}{\beta} - \beta\right).$$

We then obtain the following results:

```
exact value : 0.238
N =     100,    95% confidence interval : [0.165,0.276]
                estimated value : 0.220
N =   1 000,    95% confidence interval : [0.221,0.258]
                estimated value : 0.240
N = 10 000,     95% confidence interval : [0.232,0.244]
                estimated value : 0.238
```

We see that the approximation is much better in the case of a 'put' than for a 'call'. This is easily explained by a calculation (or estimate) of the variance. The case of a calculation of a 'call' has essentially the same characteristics as the preceding case.

In Section 1.4, we review a number of methods to reduce the variance. To illustrate the general methods we shall use the example that we have developed.

1.4   Methods to reduce the variance

We have seen that the rate of convergence of the Monte-Carlo method is of the order of $\sigma/\sqrt{n}$. There are numerous techniques (called reduction of variance techniques) to improve this method, which try to reduce the value σ^2. The general idea is to give another representation, in the form of an expected value, of the quantity to be calculated:

$$\mathbf{E}\left(X\right) = \mathbf{E}\left(Y\right),$$

trying to reduce the variance. We are going to go through several of these methods that are applicable in practically all simulations. Some techniques which are more specific to examples treated in subsequent chapters are given later.

Preferential sampling or weighting function. Assume that we want to calculate

$$\mathbf{E}(g(X))$$

and that the distribution of X is $f(x)\,dx$ (over $\mathbf{R}$ to fix ideas). The quantity that we want to evaluate therefore becomes

$$\mathbf{E}(g(X)) = \int_{\mathbf{R}} g(x)f(x)\,dx.$$

Now, let $\tilde{f}$ be the density of another distribution such that $\tilde{f} > 0$ and $\int_{\mathbf{R}} \tilde{f}(x)\,dx = 1$. It is clear that $\mathbf{E}(g(X))$ can also be written as

$$\mathbf{E}(g(X)) = \int_{\mathbf{R}} \frac{g(x)f(x)}{\tilde{f}(x)}\,\tilde{f}(x)\,dx.$$

This means that $\mathbf{E}(g(X)) = \mathbf{E}\big(g(Y)f(Y)/\tilde{f}(Y)\big)$, if Y follows the distribution $\tilde{f}(x)\,dx$ under $\mathbf{P}$. We therefore have another method of calculating

$\mathbf{E}(g(X))$ by using n trials of $Y, Y_1, \ldots, Y_n$ and by approximating $\mathbf{E}(g(X))$ by

$$\frac{1}{n}\left(\frac{g(Y_1)f(Y_1)}{\tilde{f}(Y_1)} + \cdots + \frac{g(Y_n)f(Y_n)}{\tilde{f}(Y_n)}\right).$$

If we set $Z = g(Y)f(Y)/\tilde{f}(Y)$, we shall improve the algorithm if $\mathrm{Var}(Z) < \mathrm{Var}(g(X))$. It is easy to calculate the variance of Z:

$$\mathrm{Var}(Z) = \mathbf{E}(Z^2) - \mathbf{E}(Z)^2 = \int_{\mathbf{R}} \frac{g^2(x)f^2(x)}{\tilde{f}(x)}\, dx - \mathbf{E}(g(X))^2.$$

If $g(x) > 0$, we can verify that, by taking $\tilde{f}(x) = (g(x)f(x))/(\mathbf{E}(g(X)))$ we cancel $\mathrm{Var}(Z)$! We should not attach too much importance to this result as it relies on the fact that we know $\mathbf{E}(g(X))$, and this is exactly the quantity that we want to calculate.

However, this allows us to justify the following heuristic: take $\tilde{f}(x)$ as close as possible to $|g(x)f(x)|$ then normalize it (divide by $\int \tilde{f}(x)\, dx$) to obtain a density whose distribution is easily simulated. Obviously, the constraints that we have imposed are largely contradictory and often make this a delicate exercise.

We shall give a simple example to fix ideas. Suppose that we want to calculate

$$\int_0^1 \cos\left(\pi x/2\right)\, dx.$$

This corresponds to $g(x) = \cos(x)$ and $f(x) = \mathbf{1}_{[0,1]}(x)$. We can then approximate cos by a second degree polynomial. As cos is even, and is equal to 0 at $x = 1$ and 1 at $x = 0$, it is natural to take $\tilde{f}(x)$ of the form $\lambda(1 - x^2)$. Normalizing we obtain, $\tilde{f}(x) = (1 - x^2)/3$. On calculating the variances, we establish that this method has reduced the variance by a factor of 100.

We shall show how to apply this method in the case of the calculation of a 'put' (1.2). More precisely, we want to calculate

$$P = \mathbf{E}\left(\left(1 - e^{\beta G}\right)_+\right).$$

The function $e^x - 1$ is close to x when x is not too large. This suggests taking P in the form:

$$P = \int_{\mathbf{R}} \frac{\left(1 - e^{\beta x}\right)_+}{\beta|x|}\beta|x|e^{-x^2/2}\, \frac{dx}{\sqrt{2\pi}}.$$

The change of variable, $x = \sqrt{y}$ over $\mathbf{R}^+$ and $x = -\sqrt{y}$ over $\mathbf{R}^-$, allows us to write P in the form:

$$P = \int_0^{+\infty} \frac{\left(1 - e^{\beta\sqrt{y}}\right)_+ + \left(1 - e^{-\beta\sqrt{y}}\right)_+}{\sqrt{2\pi}\sqrt{y}}e^{-y/2}\, \frac{dy}{2}.$$

If we note that $e^{-x/2}dx/2$ is the exponential distribution of a random variable Y with parameter $1/2$. We can then write:

$$P = \mathbf{E}\left(\frac{\left(1 - e^{\beta\sqrt{Y}}\right)_+ + \left(1 - e^{-\beta\sqrt{Y}}\right)_+}{\sqrt{2\pi}\sqrt{Y}} \right).$$

We can compare this with the preceding method.

```
exact value : 0.23842
N =     100,   95% confidence interval : [0.239,0.260]
               estimated value : 0.249
N =   1 000,   95% confidence interval : [0.235,0.243]
               estimated value : 0.239
N =  10 000,   95% confidence interval : [0.237,0.239]
               estimated value : 0.238
```

We see a marked improvement in the precision of the calculation, for 10 000 samples the relative error changes from 6% in the initial method to 1% thanks to this method of preferential sampling.

Control variables.   In its most simple version, we write $\mathbf{E}(f(X))$ in the form:

$$\mathbf{E}(f(X)) = \mathbf{E}(f(X) - h(X)) + \mathbf{E}(h(X)),$$

where $\mathbf{E}(h(X))$ can be calculated explicitly and $\mathrm{Var}(f(X) - h(X))$ is appreciably smaller than $\mathrm{Var}(f(X))$. We then use a Monte-Carlo method to evaluate $\mathbf{E}(f(X) - h(X))$ and direct calculation for $\mathbf{E}(h(X))$.

We start by giving a simple example. Assume that we want to calculate $\int_0^1 e^x \, dx$. Since in the neighbourhood of 0, $e^x \approx 1 + x$, we can write:

$$\int_0^1 e^x \, dx = \int_0^1 (e^x - 1 - x) \, dx + \frac{3}{2}.$$

It is easy to verify that the variance of the method then reduces appreciably.

We now give another example, by considering the price of a 'call' (1.1). It is easy to verify that the price of the 'put' and that of the 'call' satisfy the relation:

$$C - P = \mathbf{E}\left(e^{\beta G} - K\right) = e^{\beta^2/2} - K.$$

The idea is then to write $C = P + e^{\beta^2/2} - K$ and to carry out a Monte-Carlo method for P. We have already seen that the error of the method is then significantly less.

Antithetic variables.   Assume that we want to calculate:

$$I = \int_0^1 f(x) \, dx.$$

Since $x \to 1 - x$ leaves the measure dx invariant, we also have:

$$I = \frac{1}{2} \int_0^1 (f(x) + f(1 - x)) \, dx.$$

We can therefore calculate I in the following way. We choose n independent random variables $U_1, \ldots, U_n$ following the uniform distribution over $[0, 1]$, and we approximate I by

$$I_{2n} = \frac{1}{n} \left(\frac{1}{2}(f(U_1) + f(1 - U_1)) + \cdots + \frac{1}{2}(f(U_n) + f(1 - U_n)) \right)$$

$$= \frac{1}{2n} \left(f(U_1) + f(1 - U_1) + \cdots + f(U_n) + f(1 - U_n) \right).$$

When we compare this method to a direct Monte-Carlo method at the end of $2n$ trials, we can show that if the function f is monotone continuous the quality of the approximation improves.

One can generalize this kind of idea to higher dimensions and to other transformations preserving the distribution of the random variable. For example, if we want to calculate the price of a 'put' (1.2), we can use the fact that the distribution of G is identical to that of $-G$ and reduce the variance of a coefficient almost by 2.

Method of stratification. This is a method well known to statisticians and often used in surveys (see Cochran, 1977). Assume that we want to calculate I, with:

$$I = \mathbf{E}(g(X)) = \int_{\mathbf{R}^d} g(x) f(x) \, dx,$$

where X is a random variable with values in $\mathbf{R}^d$ following the distribution of $f(x) \, dx$.

We take a partition $(D_i, \ 1 \leq i \leq m)$ of $\mathbf{R}^d$. We then decompose I in the following way:

$$I = \sum_{i=1}^{m} \mathbf{E}(1_{\{X \in D_i\}} g(X)) = \sum_{i=1}^{m} \mathbf{E}(g(X) | X \in D_i) \mathbf{P}(X \in D_i).$$

When we know the numbers $p_i = \mathbf{P}(X \in D_i)$, we can use a Monte-Carlo method to estimate the integrals $I_i = \mathbf{E}(g(X) | X \in D_i)$. Assume that we approximate the integral I_i by $\tilde{I}_i$ with the help of n_i independent trials, the variance of the approximation error is given by σ_i^2 / n_i, if we denote $\sigma_i^2 = \mathrm{Var}(g(X) | X \in D_i)$. We then approximate I by $\tilde{I}$ with:

$$\tilde{I} = \sum_{i=1}^{m} p_i \tilde{I}_i.$$

Since the samples used to obtain the estimates $\tilde{I}_i$ are assumed independent, we easily show that the variance of the estimate $\tilde{I}$ becomes

$$\sum_{i=1}^{m} p_i^2 \frac{\sigma_i^2}{n_i}.$$

It is then natural to minimize this error for a fixed number of trials $\sum_{i=1}^{m} n_i = n$. We can verify that the n_i, which minimizes the variance of $\tilde{I}$, is given by

$$n_i = n \frac{p_i \sigma_i}{\sum_{i=1}^{m} p_i \sigma_i}.$$

The minimum of the variance of $\tilde{I}$ then becomes

$$\frac{1}{n} \left(\sum_{i=1}^{m} p_i \sigma_i \right)^2.$$

This is less than the variance that is obtained with n random trials by the classical Monte-Carlo method. In fact, this variance becomes

$$\text{Var}\,(g(X)) = \mathbf{E}\left(g(X)^2\right) - \mathbf{E}\left(g(X)\right)^2$$
$$= \sum_{i=1}^{m} p_i \mathbf{E}\left(g^2(X)|X \in D_i\right) - \left(\sum_{i=1}^{m} p_i \mathbf{E}\left(g(X)|X \in D_i\right) \right)^2.$$

From which by using conditional variances σ_i:

$$\text{Var}\,(g(X)) = \sum_{i=1}^{m} p_i \, \text{Var}\,(g(X)|X \in D_i) + \sum_{i=1}^{m} p_i \mathbf{E}\,(g(X)|X \in D_i)^2$$
$$- \left\{ \sum_{i=1}^{m} p_i \mathbf{E}\,(g(X)|X \in D_i) \right\}^2.$$

We then use, twice, the convexity inequality for x^2, $\left(\sum_{i=1}^{m} p_i a_i \right)^2 \leq \sum_{i=1}^{m} p_i a_i^2$ if $\sum_{i=1}^{m} p_i = 1$, to show that:

$$\text{Var}\,(g(X)) \geq \sum_{i=1}^{m} p_i \, \text{Var}\,(g(X)|X \in D_i) \geq \left(\sum_{i=1}^{m} p_i \sigma_i \right)^2.$$

This proved that, provided we have an optimal strategy of trials, we can obtain, by stratification, a reduced estimate of variance. Note however that we can seldom calculate σ_i, which limits the application of this technique (but we can always estimate them with the help of the Monte-Carlo method).

Note also that it is possible to obtain an estimate of the variance that is greater than the initial estimate if the assignment of the points to domains is arbitrary. Despite this, there exist other strategies to choose the points on domains that reduce the variance. For example, the strategy that assigns a number of points proportional to the probability of the domain:

$$n_i = n p_i.$$

We then obtain an estimate of variance equal to

$$\frac{1}{n} \sum_{i=1}^{m} p_i \sigma_i^2.$$

Now, we see that $\sum_{i=1}^{m} p_i \sigma_i^2$ is a bound for $\mathrm{Var}\,(g(X))$. This allocation strategy is sometimes useful when we explicitly know the probabilities p_i. For further results on these techniques we can consult Cochran, (1977).

Average value or conditioning.   Suppose that we want to calculate

$$\mathbf{E}(g(X,Y)) = \int g(x,y) f(x,y) \, dx \, dy,$$

where $f(x,y) \, dx \, dy$ is the distribution of the pair (X,Y). If we set:

$$h(x) = \frac{1}{m(x)} \int g(x,y) f(x,y) \, dy,$$

with $m(x) = \int f(x,y) \, dy$, it is easy to see that $\mathbf{E}(g(X,Y)) = \mathbf{E}(h(X))$. In effect, the distribution of X is $m(x) \, dx$, and therefore,

$$\mathbf{E}(h(X)) = \int m(x) h(x) \, dx = \int dx \int g(x,y) f(x,y) \, dy = \mathbf{E}(g(X,Y)).$$

We can recover this result by noting that

$$\mathbf{E}\,(g(X,Y)|X) = h(X).$$

This interpretation as a conditional expectation also allows us to prove that

$$\mathrm{Var}(h(X)) \leq \mathrm{Var}(g(X,Y)).$$

If we can explicitly calculate the function $h(x)$, it is preferable to use a Monte-Carlo method for $h(X)$.

1.5   Sequences with weak discrepancy

Another way to improve Monte-Carlo type methods is to abandon the random character of the trials and to choose the points in a 'more ordered' way. We look for deterministic sequences $(x_i, i \geq 0)$, which allow us to approximate integrals by a formula of the form:

$$\int_{[0,1]^d} f(x) \, dx \approx \lim_{n \to +\infty} \frac{1}{n} (f(x_1) + \cdots + f(x_n)).$$

In this case, we talk of the *quasi-Monte-Carlo* method. We can find sequences such that the rate of convergence of the approximation is of order $K \, (\log(n)^d / n)$, but with the condition that the function f has a certain regularity, which is significantly better than a Monte-Carlo method. It is this type of sequence that we call a sequence with weak discrepancy.

We start by giving the definition of an equidistributed sequence.

Definition 1.5.1 We say that $(x_n)_{n\geq 1}$ is an equidistributed sequence over $[0,1]^d$ if one of the following (equivalent) properties is satisfied. (If x and y are two points of $[0,1]^d$, $x \leq y$ if and only if by definition $x_i \leq y_i$, for all $1 \leq i \leq d$.)

- For all $y = (y^1, \cdots, y^d) \in [0,1]^d$:

$$\lim_{n\to+\infty} \frac{1}{n}\sum_{k=1}^{n} 1_{\{x_k \in [0,y]\}} = \prod_{i=1}^{d} y^i = \text{Volume}([0,y]),$$

  where $[0,y] = \{z \in [0,1]^d, z \leq y\}$.

- $D_n^*(x) = \displaystyle\sup_{y\in[0,1]^d} \left| \frac{1}{n}\sum_{k=1}^{n} 1_{\{x_k \in [0,y]\}} - \text{Volume}([0,y]) \right| \to 0.$

- For every Riemann integrable function f defined over $[0,1]^d$:

$$\lim_{n\to+\infty} \frac{1}{n}\sum_{k=1}^{n} f(x_k) = \int_{[0,1]^d} f(x)\,dx.$$

$D_n^*(x)$ is called the *discrepancy at the origin* of the sequence x.

Remark

- If $(U_n)_{n\geq 1}$ denotes a sequence of independent random variables with uniform distribution over $[0,1]$, the random sequences $(U_n(\omega))_{n\geq 1}$ will be almost surely equidistributed. Further, we have an iterated logarithmic distribution for the discrepancy:

$$\text{almost surely} \qquad \overline{\lim_n} \sqrt{\frac{2n}{\log(\log n)}} D_n^*(U) = 1.$$

- We say that a sequence has *weak discrepancy* if its discrepancy is asymptotically better than that of a random sequence. We can prove that the discrepancy of an infinite sequence strongly satisfies:

$$D_n^* > C_d \frac{(\log n)^{\max(d/2,1)}}{n} \qquad \text{for an infinite number of values of } n,$$

  where C_d is a constant only depending on d.

- There are many d-dimensional sequences with weak discrepancy. The best asymptotic discrepancies known are of order $((\log n)^d)/(n)$. These sequences have a quasi-optimal discrepancy considering the preceding remark.

- These sequences are asymptotically better than a sequence of random numbers. However, in practice, for values of n between 10^3 and 10^6, the discrepancies of the best-known sequences are not as good as the asymptotic results lead us to hope, particularly for dimensions higher than 10.

Another application of weak discrepancy sequences is to give an *a priori* estimate of the error made during numerical integration, for functions with finite variation, via the Koksma–Hlawka formula. As opposed to random sequences, which give confidence intervals for a given probability, this bound is effective and deterministic. We should however note that this bound is almost always far from the real value of the error and that the variation of a function is a quantity very difficult to evaluate. Proposition 1.5.2 clarifies this bound.

Proposition 1.5.2. (Koksma–Hlawka inequality)   *If g is a function of finite variation in the sense of Hardy and Krause with variation $V(g)$, then:*

$$\forall n \geq 1 \qquad \left| \frac{1}{N} \sum_{k=1}^{N} g(x_k) - \mathbf{E}(X) \right| \leq V(g) D_N^*(x).$$

Remark The general definition of a function with finite variation in the sense of Hardy and Krause is relatively complicated (see Neiderreiter, 1992). However, in one dimension, this notion coincides with that of a classical function with finite variation. Further, in d dimensions, if g is d times continuously differentiable, the variation $V(g)$ is given by

$$V(g) = \sum_{k=1}^{d} \sum_{1 \leq i_1 < \cdots < i_k \leq d} \int_{[0,1]^d} \left| \frac{\partial^k g(x)}{\partial x_{i_1} \cdots \partial x_{i_k}} \right| dx.$$

We therefore see that as the dimension increases it is increasingly 'difficult' to have finite variation. In particular, the indicator functions $(\mathbf{1}_{\{f(x_1,\ldots,x_d) > \lambda\}}$ with f regular) do not necessarily have finite variation when the dimension is greater than or equal to two.

We shall now give some examples of weak discrepancy sequences, often the most used in practice. There are also many others (see Neiderreiter, 1992 for other examples).

Van Der Corput sequences.   Let p be an integer strictly greater than 1. Let n be a positive integer we shall denote by $a_0, a_1, \ldots, a_r$ the unique p-adic decomposition satisfying:

$$n = a_0 + \cdots + a_r p^r,$$

with $0 \leq a_i < p$ for $0 \leq i \leq r$, and $a_r > 0$. The Van Der Corput sequence in base p is given by

$$\phi_p(n) = \frac{a_0}{p} + \cdots + \frac{a_r}{p^{r+1}}.$$

We can understand the definition of $\phi_p(n)$ in the following way. We write the number n in base p:

$$n = a_r a_{r-1} \ldots a_1 a_0, \quad \text{then} \quad \phi_p(n) = 0, a_0 a_1 \ldots a_r,$$

where we must understand the notation $0, a_0 a_1 \ldots a_r$ as being the p-adic decomposition of a number.

Halton sequences. Halton sequences are multidimensional generalizations of Van Der Corput sequences. Let $p_1, \ldots, p_d$ be the first d prime numbers. The Halton sequence is defined by, if n is an integer:

$$x_n^d = (\phi_{p_1}(n), \ldots, \phi_{p_d}(n)), \tag{1.3}$$

where $\phi_{p_i}(n)$ is the Van Der Corput sequence in base p_i.

The discrepancy of the d-dimensional Halton sequence is bounded by

$$D_n^* \leq \frac{1}{n} \prod_{i=1}^{d} \frac{p_i \log(p_i n)}{\log(p_i)}.$$

Faure sequence. The Faure sequence in d dimensions is defined in the following way. Let r be an odd prime number greater than d (we can take, for example, $r = 11$ in the case where $d = 8$). We then define a mapping T operating on a set of x written in the form:

$$x = \sum_{k \geq 0} \frac{a_k}{r^{k+1}},$$

the sum being finite. For one such x we then set:

$$T(x) = \sum_{k \geq 0} \frac{b_k}{r^{k+1}}$$

with $b_k = \sum_{i \geq k} C_k^i a_i \bmod r$, C_k^i being the binomial coefficients. We can then define the Faure sequence in the following way:

$$x_n = \left(\phi_r(n-1), T(\phi_r(n-1)), \ldots, T^{d-1}(\phi_r(n-1)) \right). \tag{1.4}$$

This sequence has discrepancy bounded by $C \left(\log(n)^d \right)/(n)$.

Irrational translations of the torus. These sequences are given in the form:

$$x_n = (\{n\alpha_i\})_{1 \leq i \leq d}, \tag{1.5}$$

where $\{x\}$ is the fractional part of the number x and $\alpha = (\alpha_1, \ldots, \alpha_d)$ with $(1, \alpha_1, \ldots, \alpha_d)$ a family of $\mathbf{Q}$. We can choose, for example, $\alpha = (\sqrt{p_1}, \ldots, \sqrt{p_d})$. We can prove, for this sequence, that it has a discrepancy of $o\left(1/n^{1-\epsilon}\right)$ for all $\epsilon > 0$. This sequence is used, in particular, in the commercial software NAG.

1.6   Simulation of random variables

To implement the Monte-Carlo method on a computer, we proceed in the following way. We assume we know how to construct a sequence of numbers $(U_n)_{n \geq 1}$ which is a sequence of independent uniform random variables on the interval $[0, 1]$, and we look for a function $(u_1, \ldots, u_p) \mapsto F(u_1, \ldots, u_p)$ such that the distribution of the random variable $F(U_1, \ldots, U_p)$ is the desired distribution $\mu(dx)$. The sequence of random variables $(X_n)_{n \geq 1}$ where $X_n = F(U_{(n-1)p+1}, \ldots, U_{np})$ is then a sequence of independent random variables following the desired distribution μ.

The sequence $(U_n)_{n \geq 1}$ is achieved practically by successive calls to a pseudorandom number generator. Most of the languages available in modern computers have a random function, already programmed, whose output is a pseudorandom number between 0 and 1, or a random integer in a fixed interval (this function is called `rand( )` in C ANSI, or `random` in Turbo Pascal).

Remark The function F can, in certain cases (in particular when we want to simulate a stopping time), depend on the whole sequence $(U_n)_{n \geq 1}$, and not on a fixed number of U_i. The preceding method is still useful if we know how to simulate X with the help of an almost surely finite number of U_i, this number may depend on chance. This is the case, for example, in the algorithm to simulate a random Poisson variable (see Section 1.6.2).

1.6.1   *Simulation of a uniform distribution on* $[0, 1]$

We shall show how to construct random number generators in the case where those in the available software are not entirely satisfactory.

The simplest and most commonly used method is the method of linear congruencies. We generate a sequence $(x_n)_{n \geq 0}$ of integers between 0 and $m - 1$ in the following way:

$$\begin{cases} x_0 = \text{initial value} \in \{0, 1, \ldots, m - 1\} \\ x_{n+1} = ax_n + b \, (\text{modulo } m) \end{cases}$$

a, b, m being integers which must be chosen carefully if we want the characteristic statistics of the sequence to be satisfactory.

Sedgewick (1987), recommends the following choice:

$$\begin{cases} a = 31415821 \\ b = 1 \\ m = 10^8 \end{cases}$$

This method allows us to simulate pseudorandom integers between 0 and $m - 1$; to obtain a random real number between 0 and 1 we divide the random integer generated by m.

The generator above gives acceptable results in the current cases. However, its period (here $m = 10^8$) can sometimes appear insufficient. We can obtain random number generators with arbitrarily long period by increasing m. The interested reader will find much information on random number generators and how to program them in a computer in Knuth (1981) and L'Ecuyer (1990).

1.6.2 *Simulation of other random variables*

We shall show how we can simulate, starting from a sequence of uniform random variables over $[0, 1]$, random variables simulating certain common distributions. We restrict ourselves to random variables that will occur in the rest of the book; that is, the Gaussian, exponential and Poisson random variables. Of course, we can simulate many other distributions. We will find an almost exhaustive panorama of these methods in Devroye (1986).

Simulation of Gaussian variables. A classical method to simulate Gaussian random variables starts on the observation that, if (U_1, U_2) are two independent uniform random variables over $[0, 1]$:

$$(\sqrt{-2 \log(U_1)} \cos(2\pi U_2), \sqrt{-2 \log(U_1)} \sin(2\pi U_2))$$

are a pair of independent random variables following reduced centred Gaussian distributions (i.e. of average zero and variance 1).

To simulate Gaussians of average m and variance σ, it is enough to set $X = m + \sigma g$, where g is a reduced, centred Gaussian.

Simulation of an exponential distribution. Recall that a random variable X follows an exponential distribution with parameter μ if its distribution is

$$\mathbf{1}_{\{x \geq 0\}} \mu e^{-\mu x} \, dx.$$

We can simulate X by noting that, if U follows a uniform distribution $[0, 1]$, $\log(U)/\mu$ follows an exponential distribution with parameter μ.

Simulation of a Poisson random variable. A Poisson random variable is a variable with values in $\mathbf{N}$ such that:

$$\mathbf{P}(X = n) = e^{-\lambda} \frac{\lambda^n}{n!}, \quad \text{if } n \geq 0.$$

We can prove that if $(T_i)_{i \geq 1}$ is a sequence of exponential random variables with parameter λ, then the distribution of $N_t = \sum_{n \geq 1} n \mathbf{1}_{\{T_1 + \cdots + T_n \leq t < T_1 + \cdots + T_{n+1}\}}$ is a Poisson distribution with parameter λt. N_1 therefore has the same distribution as the variable X that we want to simulate. On the other hand, we can always put the exponential variables T_i in the form $-\log(U_i)/\lambda$, where $(U_i)_{i \geq 1}$ are independent random variables following the uniform distribution over $[0, 1]$. N_1 is then written as

$$N_1 = \sum_{n \geq 1} n \mathbf{1}_{\{U_1 U_2 \ldots U_{n+1} \leq e^{-\lambda} < U_1 U_2 \ldots U_n\}}.$$

For the simulation of other distributions that we have not cited, or for other simulation methods for the distributions above, we can consult Bouleau (1986), Devroye (1986), and Bratley *et al.*, (1987).

1.7   Bibliographic comments

The reader wishing to consolidate his knowledge of probability can consult Bouleau (1986) or Breiman (1968). Many elementary works treating Monte-Carlo methods are available (Hammersley and Handscomb, 1964; Rubinstein, 1981; Kalos and Whitlock, 1986; Bratley *et al.*, 1987; Ripley, 1987). The work of Luc Devroye concerning the simulation of random variables (Devroye, 1986) is an essential reference. The weak discrepancy sequences are studied in detail in Kuipers and Neiderreiter (1974) and Neiderreiter, (1992). We can also find very many bibliographic references in Neiderreiter, (1992). The reader looking for algorithms and programs allowing us to simulate uniform random variables can consult L'Ecuyer (1990) and Press *et al.* (1992).

2

Transport equations and processes

Monte-Carlo methods for the numerical solution of partial differential equations are based on the links between stochastic processes and these equations. In this chapter, we shall give the probabilistic interpretation of *linear* transport equations (or Vlasov's equation) thanks to a class of random evolutions that are Markov processes, called 'transport processes'. The corresponding Monte-Carlo methods will be presented in detail in Chapter 3 (whereas the nonlinear Boltzmann equation and its numerical solution by Monte-Carlo methods will be studied in Chapter 4).

The transport equations serve as a starting point for a sequence of physical models; in each we must model the evolution of a system of physical particles undergoing shocks and which, between the shocks, move uniformly or accelerate.

The solution of such equations is a function u representing the population of these particles; this is a function of time t, of position x, and of velocity v (x belonging to a spatial domain $\mathcal{D}$ an open set of $\mathbf{R}^d$ and v belonging to a domain $\mathcal{V}$ which is $\mathbf{R}^d$ or part of $\mathbf{R}^d$). The transport equations that we shall consider in this chapter, and in subsequent chapters, will be linear and may be written as

$$
\begin{cases}
\dfrac{\partial u}{\partial t} + v \cdot \dfrac{\partial u}{\partial x} + \tau u = \mathcal{L}u + f, \\
u(0, \cdot) = g,
\end{cases}
$$

where

- $\tau = \tau(x, v)$;
- $\mathcal{L}$ is an integral operator with respect to the variable v, depending on the parameter x of the form:

$$
\mathcal{L}u(v) = \int_{\mathcal{V}} l\left(x, v, v'\right) u(v')\, dv';
$$

- $f = f(t, x, v)$ and $g = g(x, v)$ are the data of the problem defined over the whole space and velocity domain; these correspond to a source term and to an initial condition.

In physical terms, $u(t, x, v)$ is interpreted as a particle density, the inverse of $\tau(x, v)$ is interpreted as an average time of free flight, the quantity $|v|/\tau(x, v)$ is interpreted as the average free path of a particle at position x having velocity v (in general, $\tau(x, v)$ only depends on x and $|v|$).

The introduction of the integral operator $\mathcal{L}$ corresponds to the 'gain' part of the shock modelling. The coefficient $r(x,v) = \tau(x,v) - \int_{\mathcal{V}} l\,(x,v',v)\,dv'$ is called the damping coefficient (it is zero if the number of particles is conserved in the shock).

More generally, we can consider a problem where the particles accelerate between the shocks, in which case the equation satisfied by the particle density is the following:

$$\begin{cases} \dfrac{\partial u}{\partial t} + v \cdot \dfrac{\partial u}{\partial x} + \mathrm{div}_v(au) + \tau u = \mathcal{L}u + f, \\[2mm] u(0,\cdot) = g. \end{cases}$$

This equation is called 'Vlasov's equation' ($\mathcal{V}$ must be an open set of $\mathbf{R}^d$). The vector $a = a(x,v)$ corresponds to an acceleration term. In what follows, we shall often include the two types of equation under the term 'transport equation', although we may mean simple transport equations of the first type where a is zero.

The organization of this chapter is as follows.

We consider a differential equation satisfied by the function $X(t)$ taking values in $\mathbf{R}^d$:

$$\frac{dX(t)}{dt} = V(t),$$

where $V(t)$ is a random process. (More generally, we can consider the ordinary differential equation $dX(t)/dt = b(X(t), V(t))$ where $b : \mathbf{R}^d \times \mathcal{V} \to \mathbf{R}^d$.)

This equation can model, for example, the movement of a particle in a known force field, subject to shock at random times. We shall show that, subject to some properties of the process $V(t)$, the pair $\{X(t), V(t)\}$ is a Markov process and we shall give the form of the infinitesimal generator of the associated semigroup. Then, we shall associate this process with the 'Kolmogorov equations', which will allow us to give a probabilistic interpretation for the 'transport equations'. We first do this in a simple case where the set $\mathcal{V}$ is a discrete set. This is the object of Sections 2.2 and 2.3. After discussing a classical property of approximation of transport by diffusion (Section 2.4), we generalize, in Section 2.5, the results of Section 2.3 to the case where $\mathcal{V}$ is the space $\mathbf{R}^d$ (the case where $\mathcal{V}$ is part of $\mathbf{R}^d$ is treated in an identical manner).

To simplify the presentation in the rest of the chapter, we assume that the spatial domain for x is the whole of $\mathbf{R}^d$, in order not to be preoccupied with boundary conditions; we shall return to the equation of boundary conditions in Section 3.4.

Remark: interpretation in neutron physics. Specialists in neutron physics prefer to work, not with the function $u(x,v)$, which is interpreted as a neutron density, but with the function $\phi(x,v) = |v|^2 u(x,v)$, which we call the 'neutron flux'; nevertheless, the equation satisfied by ϕ has the same form as the preceding equation. We denote $v = |v|\Omega$, $E = |v|^2/2$, and $\phi = \phi(x; \Omega, E)$, then this equation becomes, in simple cases:

$$\frac{1}{|v|}\frac{\partial \phi}{\partial t} + \Omega.\frac{\partial \phi}{\partial x} + \sum_t (x; E)\phi(x; \Omega, E)$$

$$= \int_{S^2} \int_{R^+} \sum_s (x; \Omega', E'; \Omega, E)\phi(x; \Omega', E')\, d\Omega'\, dE' + S(x, \Omega, E),$$

where $\sum_t(x; E)$ is interpreted as a total cross-section, the coefficient $\sum_s(x; \Omega', E'; \Omega, E)$ as a scattering cross-section, and the coefficient $r(x, \Omega, E) = \sum_t(x; E) - \int\int \sum_s(x; \Omega, E; \Omega', E')\, d\Omega'\, dE'$ as an absorption cross-section. Since $d\Omega'dE' = |v'|^{-1}\, dv'$, we see that the density u satisfies:

$$\frac{\partial u}{\partial t} + v.\frac{\partial u}{\partial x} + |v| \sum_t (x; E)u(x; v)$$

$$= \int_{S^2} \int_{R^+} \frac{|v'|}{|v|} \sum_s (x; \Omega', E'; \Omega, E)u(x; v')\, dv' + S(x, \Omega, E)/|v|,$$

which is an equation of the general type described above.

Notation For every $f : \mathbf{R}^d \times \mathcal{V} \to \mathbf{R}$, and $F = (F_1, \ldots, F_d): \mathbf{R}^d \times \mathcal{V} \to \mathbf{R}^d$ we denote:

$$\nabla f(x, v) = \frac{\partial f}{\partial x}(x, v) = \left(\frac{\partial f}{\partial x_1}(x, v), \ldots, \frac{\partial f}{\partial x_d}(x, v) \right),$$

$$\text{div}_x(F) = \sum_{i=1}^{d} \frac{\partial}{\partial x_i}(F_i),$$

$C_b^0(\mathbf{R}^d \times \mathcal{V})$ is the space of functions that are continuous with respect to the first variable and bounded:

$$C_b^1(\mathbf{R}^d \times \mathcal{V}) = \{f \in C_b^0(\mathbf{R}^d \times \mathcal{V}) \text{ such that } \nabla f \in C_b^0(\mathbf{R}^d \times \mathcal{V})\}.$$

2.1 Revision of Markov processes

Given a probability space $(\Omega, \mathcal{A}, P)$, we shall denote by *stochastic process* (or *random function*) a collection $\{X(t); t \geq 0\}$ of random vectors of dimension d. For every $t \geq 0$, we therefore have a mapping:

$$\Omega \rightsquigarrow \mathbf{R}^d,$$

$$\omega \to X(t, \omega),$$

X is therefore also a mapping from $\mathbf{R}_+ \times \Omega$ to $\mathbf{R}^d$, which produces $X(t, \omega)$ for each couple (t, ω). All the processes that we consider will always satisfy the measurability property with respect to the pair (t, ω), for product sigma algebra $\mathcal{B}_+ \otimes \mathcal{A}$, where $\mathcal{B}_+$ denotes the Borel sigma algebra of $\mathbf{R}_+$.

We denote by the path of the process X the mapping:

$$t \to X(t, \omega)$$

from $\mathbf{R}_+$ into $\mathbf{R}^d$, ω being fixed. The set of paths is therefore a collection indexed by ω of mappings from $\mathbf{R}_+$ into $\mathbf{R}^d$.

We consider a measurable set G equipped with a sigma algebra $\mathcal{G}$. For example, G may be a finite or countable set or G may be $\mathbf{R}^d$ (then $\mathcal{G}$ is the sigma algebra $\mathcal{B}_d$ of Borel sets of $\mathbf{R}^d$).

We shall now give the definition of a Markov process with values in G.

Definition 2.1.1 A stochastic process $\{Z(t); t \geq 0\}$ with values in a measurable space $(G, \mathcal{G})$ will be called *Markovian* (or called a '*Markov process*') if for all $0 < s < t$, and for:

$$0 = t_0 < t_1 < \cdots < t_n < s \quad \text{and} \quad z_0, z_1, \ldots, z_n, z \in G,$$

we have

$$\mathbf{P}(Z(t) \in B / Z(0) = z_0, Z(t_1) = z_1, \ldots, Z(t_n) = z_n, Z(s) = z)$$
$$= \mathbf{P}(Z(t) \in B / Z(s) = z) \quad \forall B \in \mathcal{G}.$$

Remark The terms on the left (resp. right) hand side are functions of $z_0, z_1, \ldots, z_n, z$ (resp. of z) almost surely equal for the law of $(Z(0), Z(t_1), \ldots, Z(t_n), Z(s))$ (resp. of $Z(s)$). We cannot replace '$Z(t) \in B$' by a condition of the type '$Z(t) = z'$' except if G is a finite or countable set in which case the preceding definition is equal to

$$\mathbf{P}(Z(t) = z' / Z(0) = z_0, Z(t_1) = z_1, \ldots, Z(t_n) = z_n, Z(s) = z)$$
$$= \mathbf{P}(Z(t) = z' / Z(s) = z) \quad \forall z' \in G.$$

Definition 2.1.2 We say that the Markov process $\{Z(t), t \geq 0\}$ is *homogeneous* if the quantity $\mathbf{P}(Z(t) \in B / Z(s) = z)$ only depends on s and t through the difference $t - s$.

In what follows, we are only interested in homogeneous Markov processes. For every bounded measurable function f we shall write:

$$\mathbf{E}_z\left(f(Z(t))\right) = \mathbf{E}\left(f(Z(t)) / Z(0) = z\right)$$

and also

$$\mathbf{P}_z\left(\cdot\right) = \mathbf{P}\left(\cdot/Z(0) = z\right).$$

From the definitions above, we see that for every positive t and s, we have:

$$\mathbf{P}(Z(t+s) \in B/Z(s)) = \mathbf{P}_{Z(s)}(Z(t) \in B).$$

2.1.1 Semigroup associated with a Markov process

With every homogeneous Markov process $\{Z(t), t \geq 0\}$, we can associate a transition semigroup $\{Q_t, t \geq 0\}$ defined as follows. For every $t \geq 0$, Q_t is a mapping from $G \times \mathcal{G}$ into $[0, 1]$ such that for all $z \in G$, $B \in \mathcal{G}$:

$$Q_t(z; B) = \mathbf{P}_z(Z(t) \in B).$$

We verify that $z \to Q_t(z; B)$ is measurable, and it is clear that for every $z \in G$, $B \to Q_t(z, B)$ is a probability measure. We also denote by Q_t, the linear operator, which, to every bounded measurable function $f : G \to \mathbf{R}$, associates the function $Q_t f$ defined by

$$Q_t f(z) = \int_G f(z') Q_t(z; dz');$$

which is again written as

$$Q_t f(z) = E_z\left(f(Z(t))\right).$$

From the Markov property, we see that

$$Q_{t+s}(z; B) = \mathbf{P}_z[\mathbf{P}(Z(t+s) \in B/Z(s))] = \mathbf{P}_z[\mathbf{P}_{Z(s)}(Z(t) \in B)],$$

we therefore deduce:

Proposition 2.1.3 *For all $s, t \geq 0$, we have the semigroup property:*

$$Q_{t+s}(z, B) = \int_G Q_t(z', B) Q_s(z, dz')$$

for every B in $\mathcal{G}$, and

$$Q_{t+s} f = Q_t(Q_s f)$$

for every measurable bounded function f.

The *infinitesimal generator* of the semigroup Q_t (and of the associated Markov process) is the operator A, which is the derivative at the origin of the semigroup Q_t (this operator A is, in general, unbounded):

$$Af(z) = \lim_{h \downarrow 0} \frac{1}{h} \left\{ \mathbf{E}_z f(Z(h)) - f(z) \right\}.$$

The domain of A is the space of all bounded measurable functions f such that the limit above exists for all z.

2.2   Transport processes with discrete velocities

Transport processes are elementary random evolutions; in order to construct these random evolutions, we initially consider, for simplicity, the case where the velocities take their values in a finite or countable state space $\mathcal{V} = E$.

2.2.1   *Jump Markov processes*

It is useful first to construct and study the jump Markov processes with values in a finite or countable space $\mathcal{V} = E$. One such process $\{V(t), t \geq 0\}$ has trajectories that are continuous to the right and constant between jumps. The jumps are produced at random moments $0 \leq T_1(\omega) \leq T_2(\omega) \leq \cdots \leq T_n(\omega)$. If we denote by $\xi_n(\omega)$ the position of $V(t)$ just after the nth jump $T_n(\omega), n \geq 1$, the data of $\{V(t); t \geq 0\}$ is equivalent to that of the double sequence $\{T_n, \xi_n; n \geq 0\}$ (the paths of $V(t)$ are constant between the times T_n).

For certain applications, it is useful to be able to produce absorbing states: $x \in E$ is called absorbent if $V_{T_n}(\omega) = x \Rightarrow T_{n+1}(\omega) = +\infty$. We therefore assume that the times of jumps form an increasing sequence:

$$0 = T_0 < T_1 \leq T_2 \leq \cdots \leq T_n \leq \cdots$$

with $T_n \in \mathbf{R}_+ \cup \{+\infty\}$, and

$$T_n(\omega) < T_{n+1}(\omega) \quad \text{if} \quad T_n(\omega) < \infty,$$

and in addition that $T_n(\omega) \to +\infty$ as $n \to \infty$. A random function $\{V(t); t \geq 0\}$ with values in E is called a random function of jumps if it has the form:

$$V(t, \omega) = \sum_{\{n \geq 0; T_n(\omega) < \infty\}} \xi_n(\omega) \mathbf{1}_{[T_n(\omega), T_{n+1}(\omega)[}(t),$$

where the random variables ξ_n take their values in E.

We shall construct a particular random function of jumps. For this, it is useful to have a positive bounded function λ defined over E and a Markov matrix $\{\Pi(v, w); v, w \in E\}$ (that is, a matrix satisfying: $\Pi(v, w) \geq 0$, for all v, w and $\sum_w \Pi(v, w) = 1$ for all v); this is a transition matrix of a Markov chain in discrete time with values in E.

The random variables T_1 and ξ_1 are conditionally independent, given ξ_0. The conditional law T_1 given ξ_0 is an exponential law with parameter $\lambda(x_0)$ and the conditional law of ξ_1 given ξ_0 is given by $\Pi(x_0, .)$.

More generally, for all $n \geq 1$, $T_{n+1} - T_n$ and ξ_{n+1} are conditionally independent given (ξ_n, T_n), the conditional law of $(T_{n+1} - T_n)$ given (T_n, ξ_n) is an exponential law with parameter $\lambda(\xi_n)$ and the conditional law of ξ_{n+1} is given by

$$\Pi(\xi_n, \cdot).$$

What we have done above exactly complements the conditional law of the infinite sequence $\{(T_n, \xi_n), n \geq 1\}$ given ξ_0, and therefore also the conditional law of $\{V(t), t > 0\}$ given $V(0)$.

We then verify that the random function of jumps $\{V(t), t \geq 0\}$ with values in E is a Markov process called a *jump Markov process* (or Markov chain in continuous time), and this process is homogeneous, that is,

$$\mathbf{P}\left(V(t) = w / V(s) = v\right) = Q_{t-s}(v, w)$$

for all $t > s$. Q_t is a 'Markov matrix' over E, called the transition matrix at time t. It is the semigroup associated with the homogeneous Markov process $\{V(t), t \geq 0\}$. We shall denote below by μ_t the probability law of $V(t)$ over E, $t \geq 0$. μ_0 is called the 'initial law' of the process $\{V(t); t \geq 0\}$.

Proposition 2.2.1 *Let $\{V(t), t \geq 0\}$ be a jump Markov process, with initial law μ and with transition matrices $\{Q_t, t > 0\}$. For all $n \in \mathbb{N}$, $0 < t_1 < \cdots < t_n$ and $v_0, v_1, \ldots, v_n \in E$, the law of the random vector $(V(0), V(t_1), \ldots, V(t_n))$ is given by*

$$P\left(V(0) = v_0, V(t_1) = v_1, V(t_2) = v_2, \ldots, V(t_n) = v_n\right)$$
$$= \mu_0(v_0) Q_{t_1}(v_0, v_1) Q_{t_2 - t_1}(v_1, v_2) \cdots Q_{t_n - t_{n-1}}(v_{n-1}, v_n).$$

Consequently, for all $t > 0$:

$$\mu_t = \mu_0 \, Q_t,$$

in the sense where $\mu_t(w) = \sum_{v \in E} \mu_0(v) \, Q_t(v, w)$, and for every positive bounded function $g : E \to \mathbf{R}$,

$$\mathbf{E}_v\{g(V(t))\} = (Q_t \, g)(v)$$
$$= \sum_{w \in E} Q_t(v, w) g(w).$$

Moreover, the transition matrices $\{Q_t, t > 0\}$ satisfy the semigroup relation (the Chapman–Kolmogorov equation):

$$Q_s Q_t = Q_{s+t},$$

in the sense of $\sum_{z \in E} Q_s(v, z) Q_t(z, w) = Q_{s+t}(v, w)$ for $s, t > 0$ and $v, w \in E$.

Proof It follows immediately from the definition of conditional probabilities and from the Markov property that

$$\mathbf{P}(V(0) = v_0, V(t_1) = v_1, V(t_2) = v_2, \ldots, V(t_n) = v_n)$$
$$= \mathbf{P}\left(V(0) = v_0\right)\mathbf{P}(V(t_1) = v_1/V(0) = v_0)$$
$$\times \mathbf{P}\left(V(t_2) = v_2/V(0) = v_0, V(t_1) = v_1\right) \times \cdots$$
$$\times \mathbf{P}\left(V(t_n) = v_n/V(0) = v_0, V(t_1) = v_1, \ldots, V(t_{n-1}) = v_{n-1}\right)$$
$$= \mathbf{P}\left(V(0) = v_0\right)\mathbf{P}(V(t_1) = v_1/V(0) = v_0)$$
$$\times \mathbf{P}\left(V(t_2) = v_2/V(t_1) = v_1\right) \times \cdots \times \mathbf{P}\left(V(t_n) = v_n/V(t_{n-1}) = v_{n-1}\right)$$
$$= \mu(v_0)Q_{t_1}(v_0, v_1)Q_{t_2-t_1}(v_1, v_2) \times \cdots \times Q_{t_n-t_{n-1}}(v_{n-1}, v_n).$$

In the case $n = 1$, this formula is written as

$$P\left(V(0) = v, V(t) = w\right) = \mu_0(v)Q_t(v, w)$$

and the second is deduced by summing over $v \in E$. By the definition of Q_t,

$$P\left(V(t) = w/V(0) = v\right) = Q_t(v, w),$$

the third result is deduced by multiplying by $g(w)$ and summing over $w \in E$. Finally, the formula above in the case $n = 2$ gives, after division by $\mu_0(x_0)$,

$$P\left(V(s) = z, V(s + t) = w/V(0) = v\right) = Q_s(v, z)Q_t(z, w),$$

the last result is deduced by summing over $z \in E$. $\square$

We have clarified the general semigroup property stated in Section 2.1.1. We shall now present some examples of jump Markov processes.

Example 1 A *Poisson process* with intensity λ is a jump Markov process of the form:

$$V(t) = \sum_{n \geq 0} n\mathbf{1}_{[T_n(\omega), T_{n+1}(\omega)[}(t),$$

where $\{T_{n+1} - T_n; n \in \mathbb{N}\}$ are independent random variables with exponential law with constant parameter λ. The associated transition matrix is

$$Q_t(v, w) = \begin{cases} e^{-\lambda t}(\lambda t)^{w-v}/(w - v)!, & \text{if } w \geq v, \\ 0, & \text{otherwise.} \end{cases}$$

Example 2 *Telegraph process.* Given a Poisson process $\{N(t)\}$ with intensity λ, and a random variable $V(0)$ with values in $E = \{-1, +1\}$ independent of $\{N(t); t \geq 0\}$, we set:

$$V(t) = V(0)(-1)^{N(t)}, \quad t \geq 0$$

$\{V(t), t \geq 0\}$ is a Markov process with transition matrix:

$$Q_t(+1, +1) = Q_t(-1, -1) = e^{-\lambda t}\sum_{n \geq 0}(\lambda t)^{2n}/(2n)!$$

$$Q_t(-1, +1) = Q_t(+1, -1) = e^{-\lambda t}\sum_{n \geq 0}(\lambda t)^{2n+1}/(2n + 1)!$$

Example 3 Let $(N(t), t \geq 0)$ be a Poisson process with intensity λ. We denote the times of jumps by $0 < T_1 < T_2 < T_3 < \cdots < T_n < \cdots$. We are further given a Markov chain in discrete time $(\xi_n, n \in \mathbf{N})$ with values in E, and transition matrix $\{\Pi(u,v); \; u, v \in E\}$, independent of the process N. We easily verify that

$$V(t) = \sum_{n \geq 0} \xi_n \mathbf{1}_{[T_n, T_{n+1}[}(t) \quad \text{for } t \geq 0,$$

is a jump Markov process, with transition matrix Q_t with:

$$Q_t(u,v) = e^{-\lambda t} \sum_{n \geq 0} \frac{\lambda^n}{n!} \Pi^n(u,v).$$

As was indicated in Section 2.1.1, with the semigroup of operators Q_t, we can associate an *infinitesimal generator* (which is the derivative to the right of Q_t at $t = 0$). Here, we have the following theorem.

Theorem 2.2.2 *Let $\{Q_t, t > 0\}$ be the semigroup of transition matrices of a jump Markov process $\{V(t), t \geq 0\}$. Then, there exists a matrix $\{A(v,w); v, w \in E\}$ satisfying:*

1. $A(v, w) \geq 0$ *if* $v \neq w$,
2. $A(v, v) = -\sum_{w \in E \setminus \{v\}} A(v, w) \leq 0$

(this last inequality being strict except if the state v is absorbent which is the infinitesimal generator of the semigroup $(Q_t))$, that is to say that we have, as h tends to 0:

$$Q_h(v, w) = hA(v, w) + \circ(h) \quad \text{if } v \neq w,$$
$$Q_h(v, v) = 1 + hA(v, v) + \circ(h).$$

Moreover, conditionally in $V(0) = v$, the time of the first jump T_1 and the position after the first jump $\xi_1 = V(T_1)$ are independent, T_1 has exponential law with parameter $q(v) = -A(v, v)$, and ξ_1 has law over E given by $A(v, \cdot)/q(v)$.

That is, $\Pi(v, w) = A(v, w)/q(v)$, $w \neq v$.

Proof We remark first that

$$\{T_1 > nh\} \subset \{V(0) = V(h) = \cdots = V(nh)\} \subset \{T_1 > nh\} \cup \{T_2 - T_1 \leq h\}.$$

Since $\mathbf{P}(T_2 - T_1 \leq h) \to 0$ when $h \to 0$, we see that if $h \to 0$, $nh \to t$ (with $nh \geq t$):

$$\mathbf{P}(T_1 > t/V(0) = v) = \mathbf{P}(V(0) = V(h) = \cdots = V(nh)/V(0) = v)$$
$$= \lim[Q_h(v, v)]^n.$$

The existence of this last limit follows from:

$$\frac{1}{h}[1 - Q_h(v,v)] \to q(v) \in [0, +\infty],$$

when $h \to 0$, and therefore

$$\mathbf{P}(T_1 > t/V(0) = v) = e^{-q(v)t}.$$

From which $q(v) < \infty$ and $q(v) = 0$ if and only if v is absorbent. We set $A(v,v) = -q(v)$. The proof of the existence of limits of $(1/h)Q_h(v,w)$ for $v \neq w$ is analogous:

$$\{T_1 \leq t, \xi_0 = v, \xi_1 = w\}$$
$$= \lim_{h \to 0, nh \to t} \sum_{1 \leq m \leq n} \{V(0) = V(h) = \cdots = V((m-1)h) = v, V(mh) = w\},$$

$$\mathbf{P}(T_1 \leq t, \xi_1 = w/V(0) = v)$$
$$= \lim \frac{1 - Q_h(v,v)^n}{1 - Q_h(v,v)} Q_h(v,w)$$
$$= \frac{1 - e^{-q(v)t}}{q(v)} \lim \frac{1}{h} Q_h(v,w).$$

Therefore, $A(v,w) = \lim(1/h)Q_h(v,w)$ exists for $v \neq w$ and

$$\mathbf{P}(T_1 \leq t, \xi_1 = w/V(0) = v) = (1 - e^{-q(v)t})\frac{A(v,w)}{q(v)},$$

from which

$$\mathbf{P}(T_1 \leq t, \xi_1 = w/V(0) = v) = \mathbf{P}(T_1 \leq t, /V(0) = v)\mathbf{P}(\xi_1 = w/V(0) = v)$$

and

$$\mathbf{P}(\xi_1 = w/V(0) = v) = \frac{A(v,w)}{q(v)}.$$

$\square$

In the case where the cardinality $E < \infty$, we immediately deduce the following result from Theorem 2.2.2, which remains true if $E = +\infty$:

Corollary 2.2.3

1. *The matrix function $\{Q_t, t \geq 0\}$ is the unique solution of the equation:*

$$\frac{dQ_t}{dt} = AQ_t, \quad t > 0; \quad Q_0 = I$$

and $u(t, v) = \mathbf{E}_v[g(V(t))]$ satisfies the 'retrograde' Kolmogorov equation:

$$\begin{cases} \dfrac{\partial u}{\partial t}(t, v) = \sum_{w \in E} A(v, w)u(t, w), \quad t > 0, \quad v \in E; \\ u(0, v) = g(v), v \in E. \end{cases}$$

2. *The matrix function $\{Q_t, t \geq 0\}$ is also the unique solution of the equation:*

$$\frac{dQ_t}{dt} = Q_t\, A, \quad t > 0, \quad Q_0 = I.$$

Moreover, the family of marginal probability laws $\{\mu_t, t \geq 0\}$ of random variables $\{V(t); t \geq 0\}$ satisfies the equation called the progressive Kolmogorov equation:

$$\frac{\partial \mu_t(v)}{\partial t} = \sum_{y \in E} \mu_t(y)A(y, v), \quad t > 0, \quad v \in E.$$

The progressive Kolmogorov equation is also called the Fokker–Planck equation and in what follows we shall often use this latter name. (Do not confuse the general concept of Fokker–Planck equations related to any Markov process with the Fokker–Planck–Landau operators that occur in the kinetic theory of plasmas.)

Proof *Point 1.* To establish the matrix retrograde Kolmogorov equation, it is sufficient to differentiate $Q_t(v, w)$ by using the semigroup property in the form:

$$Q_{t+h} = Q_h Q_t.$$

The equation for u is then deduced from the equation obtained by multiplying to the right by the column vector $\{g(\cdot)\}$.

Point 2. The matrix equation is deduced starting from the formula:

$$Q_{t+h} = Q_t Q_h.$$

The progressive Kolmogorov equation is then deduced by multiplying on the left by the row vector $\{\mu_0(.)\}$, after integration. $\qquad\qquad$ □

2.2.2 Transport processes

Let $b : \mathbf{R}^d \times E \to \mathbf{R}^d$ belong to the space $C^1(\mathbf{R}^d \times E)$. It follows from classical theorems on ordinary differential equations that for all $x \in \mathbf{R}^d$, $v \in E$, the equation

$$\begin{cases} \dfrac{dX(t)}{dt} = b(X(t), v) \\[2mm] X(0) = x \end{cases} \tag{2.1}$$

has a unique solution $X(t) = \phi_{x,v}(t)$, $t \geq 0$. We can then establish:

Theorem 2.2.4 *Given a jump Markov process $\{V(t); t \geq 0\}$ with values in E and a random vector $X_0 \in \mathbf{R}^d$, independent of $\{V(t); t \geq 0\}$, there exists a unique process $\{X(t); t \geq 0\}$ with continuous paths, such that for all ω in Ω,*

$$\begin{cases} \dfrac{dX(t, \omega)}{dt} = b(X(t, \omega), \quad V(t, \omega)), \quad t \geq 0, \\[2mm] X(0, \omega) = X_0(\omega). \end{cases} \tag{2.2}$$

Proof We use the representation

$$V(t, \omega) = \sum_{n \geq 0} \xi_n(\omega) \mathbf{1}_{[T_n(\omega), T_{n+1}(\omega)[}(t).$$

We see that the equation can be solved successively on the intervals $[T_n(\omega), T_{n+1}(\omega)[$, as follows. For $t \in [0, T_1(\omega)[$, we have:

$$X(t, \omega) = \phi_{X_0(\omega), \xi_0(\omega)}(t)$$

and we set:

$$\eta_1(\omega) = \phi_{X_0(\omega), \xi_0(\omega)}(T_1(\omega)).$$

Then, for $t \in [T_1(\omega), T_2(\omega)[$ we set:

$$X(t, \omega) = \phi_{\eta_1(\omega), \xi_1(\omega)}(t - T_1(\omega))$$

and

$$\eta_2(\omega) = \phi_{\eta_1(\omega), \xi_1(\omega)}(T_2(\omega) - T_1(\omega)),$$

and so on. It is easy to verify that we can construct the unique solution of the equation above. □

We can now show that the pair $\{(X(t), V(t)); t \geq 0\}$ (taking values in the uncountable set $\mathbf{R}^d \times E$) is a Markov process, even though $\{X(t); t \geq 0\}$ *is not Markovian.* We have:

Proposition 2.2.5 *The process $\{(X(t), V(t)), t \geq 0\}$ defined in Theorem 2.2.4 is a Markov process.*

Proof Conditionally in $(X(s), V(s)) = (x, v)$, $(X(s + u), V(s + u))$ is a function of the process $(V(s+r), 0 \leq r \leq u)$ conditioned by $V(s) = v$, which is independent jointly of $X(0)$ and of $(V(t), 0 \leq t \leq s)$ [and therefore of $((X(t), V(t)), 0 \leq t \leq s)$] since it is independent of $(V(t), 0 \leq t \leq s)$, and $X(0)$ is independent of $(V(t); 0 \leq t \leq s + u)$. $\quad\square$

This process $(X(t), V(t))$ is a particular transport process corresponding to a discrete set of velocities (general transport processes will be introduced in Section 2.5).

2.2.3 *Infinitesimal generator of the associated semigroup*

To the Markov process $\{(X(t), V(t)), t \geq 0\}$ with values in $\mathbf{R}^d \times E$ which we construct, we associate, as in Section 2.1, the linear operator $\mathbf{Q}_t$ which, to every measurable bounded function $f : \mathbf{R}^d \times E \to \mathbf{R}$, associates the function $\mathbf{Q}_t f$ defined by

$$\mathbf{Q}_t f(x, v) = \mathbf{E}_{x,v} f(X(t), V(t)).$$

Likewise, we define the infinitesimal generator $\mathcal{A}$ of the semigroup $\mathbf{Q}_t$ (and of the associated Markov process) by

$$\mathcal{A}f(x, v) = \lim_{h \downarrow 0} \frac{1}{h} \left\{ \mathbf{E}_{x,v} f(X(h), V(h)) - f(x, v) \right\}.$$

Here, this infinitesimal generator has a simple expression, which we shall now give. Recall that the infinitesimal generator of the jump Markov process $\{V(t); t \geq 0\}$ is $(A(v, v'); v, v' \in E)$.

Proposition 2.2.6 *The infinitesimal generator $\mathcal{A}$ of the Markov process $\{(X(t), V(t)), t \geq 0\}$ acts as a function f of $C_b^1(\mathbf{R}^d \times E)$:*

$$\mathcal{A}f(x, v) = b(x, v) \cdot \nabla f(x, v) + \sum_{v' \in E} A(v, v') f(x, v').$$

Proof We shall denote by $E_{x,v}$ the conditional expectation given $(X(0), V(0)) = (x, v)$. Let $f \in C_b^1(\mathbf{R}^d \times E)$, we denote by T_1 the time of the first jump of $\{V(t), t \geq 0\}$:

$$\frac{1}{h} \left\{ \mathbf{E}_{x,v} f(X(h), V(h)) - f(x, v) \right\}$$

$$= \frac{1}{h} \mathbf{E}_{x,v} \left\{ (f(X(h), V(h)) - f(x, V(h))) \, \mathbf{1}_{\{T_1 > h\}} \right\}$$

$$+ \frac{1}{h} \mathbf{E}_{x,v} \left\{ (f(X(h), V(h)) - f(x, V(h))) \, \mathbf{1}_{\{T_1 \leq h\}} \right\}$$

$$+ \frac{1}{h} \mathbf{E}_{x,v} \left\{ f(x, V(h)) - f(x, v) \right\}.$$

We easily deduce from Theorem 2.2.2 that the last term on the right-hand side converges, as h tends to 0, to $\sum_{v' \neq v} A(v, v') f(x, v')$. The second term is bounded in absolute value by

$$\sup_{x,v} |\nabla f(x, v)| \times h^{-1} \left(\mathbf{E}_{x,v}(|X(h) - x|^2) \right)^{1/2} \left(P_{x,v}(T_1 \leq h) \right)^{1/2}$$

$$\leq C \left(P_{x,v}(T_1 \leq h) \right)^{1/2},$$

which tends to zero as $h \to 0$. Finally,

$$\frac{1}{h} \mathbf{E}_{x,v} \left\{ (f(X(h), V(h)) - f(x, V(h))) \mathbf{1}_{\{T_1 > h\}} \right\}$$

$$= \frac{1}{h} \left\{ f(\phi_{x,v}(h), v) - f(x, v) \right\} P_{x,v}(T_1 > h)$$

$$= \frac{1}{h} \int_0^h b(\phi_{x,v}(s), v) \cdot \nabla f(\phi_{x,v}(s), v) \, ds \times e^{A(x,x)h}.$$

This last term converges, as h tends to 0, to $b(x, v) \cdot \nabla f(x, v)$. $\Box$

2.3   Associated Kolmogorov equations

We shall now establish the Kolmogorov equations associated with the transport process $\{(X(t), V(t)); t \geq 0\}$. For this we shall make the hypothesis:

$$\sup_{v \in E} |A(v, v)| < \infty.$$

This hypothesis is certainly satisfied in the case card $E < \infty$. It is too restrictive for many jump Markov processes, but we shall be content to study this case.

We shall use the following notation:

$$\mathbf{E}_{s,x,v}(\cdot) = \mathbf{E}(\cdot / X(s) = x, V(s) = v),$$

$$\mathbf{E}_{s,\psi}(\cdot) = \sum_{v \in E} \int_{\mathbf{R}^d} \mathbf{E}_{s,x,v}(\cdot) \psi(x, v) \, dx$$

for every probability density ψ over $\mathbf{R}^d \times E$. We shall also denote:

$$C_b^1(\mathbf{R}_+ \times \mathbf{R}^d \times E) = C^1([0, +\infty); C_b^1(\mathbf{R}^d \times E)).$$

2.3.1   Fokker–Planck equation

Following what we have already discussed, we shall use this name as a synonym for the progressive Kolmogorov equation. It follows from the calculation made in Proposition 2.2.6 that we have:

Lemma 2.3.1 *If* $f \in C_b^1(\mathbf{R}_+ \times \mathbf{R}^d \times E)$, *the function* $t \rightarrow$ $\mathbf{E}_{x,v} f(t, X(t), V(t))$ *is continuously differentiable for all* $t > 0$,

$$\frac{d}{dt}[\mathbf{E}_{x,v} f(t, \ X(t), \ V(t))] = \mathbf{E}_{x,v} \left\{ \frac{\partial f}{\partial t}(t, X(t), V(t)) + \mathcal{A}f(t, X(t), V(t)) \right\}.$$

Let μ_0 be the probability law of $(X(0), V(0))$ for all $t > 0$, we denote

$$\mu_t : \mathcal{B}_d \times E \rightarrow [0, 1]$$

the probability law of $(X(t), V(t))$, that is,

$$\mu_t(B, v) := \mathbf{P}(X(t) \in B, V(t) = v).$$

For $f \in C_b(\mathbf{R}^d \times E)$, we shall denote

$$\mu_t(f) = \sum_{v \in E} \int_{\mathbf{R}^d} f(x, v) \mu_t (dx, v)$$
$$= \mathbf{E}_{0, \mu_0} (f(X(t), V(t))).$$

It follows immediately from Lemma 2.3.1 that the collection of probability measures $\{\mu_t, t \geq 0\}$ satisfies the Fokker–Planck equation:

Proposition 2.3.2 *For all* f *in* $C_b^1(\mathbf{R}^d \times E)$, $t > 0$:

$$\mu_t(f) = \mu_0(f) + \int_0^t \mu_s(\mathcal{A}f) \, ds.$$

It is easy to see that if $X(0) = x$, the law $X(t)$ is not absolutely continuous with respect to the Lebesgue measure, since $\mathbf{P}(X(t) = \phi_{x,v}(t)) \geq \mathbf{P}(V(0) = v, T_1 > t)$. Conversely, it is not difficult to show that if the law of $X(0)$ has density $p_0(x)$ with respect to the Lebesgue measure over $\mathbf{R}^d$, then the law of $X(t)$ is absolutely continuous with respect to this Lebesgue measure, that is,

$$\mu_t(dx, v) = p(t, x, v) \, dx.$$

We denote by $\mathcal{A}^*$ the adjoint operator of $\mathcal{A}$, that is,

$$\mathcal{A}^* f(x, v) = -\mathrm{div}_x(bf)(x, v) + \sum_{v' \in E} A(v', v) f(x, v').$$

We then have:

Corollary 2.3.3 *Assume that the law of* $(X(0), V(0))$ *is of the form:*

$$\mathbf{P}(X(0) \in B, V(0) = v) = \left(\int_B p_0(x) \, dx \right) \mathbf{P}(V(0) = v), \quad B \in \mathcal{B}_d, \ v \in E,$$

then the density of the law of $\{(X(t), V(t)), t \geq 0\}$ *satisfies in the sense of distributions:*

$$\begin{cases} \dfrac{\partial p}{\partial t} = \mathcal{A}^* p, t > 0, \\[2mm] p(0, x, v) = p_0(x) \mathbf{P}(V(0) = v). \end{cases}$$

2.3.2 The retrograde Kolmogorov equation

We can now establish the 'retrograde Kolmogorov equation' associated with the Markov process $\{(X(t), V(t)), t \geq 0\}$.

Theorem 2.3.4 *Let* $g \in C_b^1(\mathbf{R}^d \times E)$. *Then,* $u(t, x, v) := \mathbf{E}_{x,v}g(X(t), V(t))$ *is the unique element of* $C_b^1(\mathbf{R}_+ \times \mathbf{R}^d \times E)$, *which is the solution of the retrograde Kolmogorov equation:*

$$
\begin{cases}
\dfrac{\partial u}{\partial t}(t, x, v) = (\mathcal{A}u)(t, x, v), \quad t > 0, \ x \in \mathbf{R}^d, \ v \in E; \\[2ex]
u(0, x, v) = g(x, v), \quad x \in \mathbf{R}^d, \ v \in E.
\end{cases}
$$

Proof Let $u \in C_b^1(\mathbf{R}_+ \times \mathbf{R}^d \times E)$ be a solution of the equation above. We set $\overline{u}(s, x, v) = u(t - s, x, v)$, $s \in [0, t]$, $x \in \mathbf{R}^d$, $v \in E$. We then remark that

$$
\frac{\partial \overline{u}}{\partial s} + \mathcal{A}\overline{u} = 0.
$$

It then follows from Lemma 2.3.1 that $\mathbf{E}_{x,v}\overline{u}(s, X(s), V(s))$ is constant for $s \in [0, t]$, therefore

$$
\begin{aligned}
\mathbf{E}_{x,v}\overline{u}(0, X(0), V(0)), &= \mathbf{E}_{x,v}\overline{u}(t, X(t), V(t)), \\
u(t, x, v), &= \mathbf{E}_{x,v}g(X(t), V(t)).
\end{aligned}
$$

Conversely, let u be given by the stated formula. Assume for the moment that we have established the regularity result $u \in C_b^1(\mathbf{R}_+ \times \mathbf{R}^d \times E)$. Fix $t > 0$, and set, for $0 \leq s \leq t$,

$$
\begin{aligned}
\overline{u}(s, x, v) &= u(t - s, x, v) \\
&= \mathbf{E}_{x,v}g(X(t - s), V(t - s)) \\
&= \mathbf{E}_{s,x,v}g(X(t), V(t)),
\end{aligned}
$$

from the homogeneity of the Markov process $\{(X(t), V(t))\}$. Let $0 \leq r < t$, then it follows from the Markov property that

$$
s \rightarrow \mathbf{E}_{r,x,v}\overline{u}(s, X(s), V(s))
$$

is constant on the interval $[r, t]$. But from Lemma 2.3.1, for $s \in]r, t[$,

$$
0 = \frac{d}{ds}\mathbf{E}_{r,x,v}[\overline{u}(s, X(s), V(s))] = \mathbf{E}_{r,x,v}\left\{\left(\frac{\partial v}{\partial s} + \mathcal{A}v\right)(s, X(s), V(s))\right\}.
$$

Passing to the limit as $r \uparrow s$, we deduce:

$$
\left(\frac{\partial \overline{u}}{\partial s} + \mathcal{A}\overline{u}\right)(s, x, v) = 0,
$$

from where

$$\frac{\partial u}{\partial t}(t, x, v) = \mathcal{A}u(t, x, v).$$

It remains to show that $u \in C_b^1(\mathbf{R}_+ \times \mathbf{R}^d \times E)$. The differentiability in t is a consequence of Lemma 2.3.1. The differentiability in x is shown with the help of classical results of the differentiability of the solution of an ordinary differential equation, and since g and its gradient are bounded we deduce that u and its gradient are bounded. $\qquad\square$

2.3.3 *Generalization*

We shall now give probabilistic interpretations for some of the classes of equations that generalize the retrograde Kolmogorov equations and the Fokker–Planck equation. In what follows, we denote by c a function belonging to $C_b^0(\mathbf{R}^d \times E)$. Note firstly the following immediate generalization of Lemma 2.3.1:

Lemma 2.3.5 *If $f \in C_b^1(\mathbf{R}_+ \times \mathbf{R}^d \times E)$, the function:*

$$t \rightarrow \mathbf{E}_{x,v} \left\{ \exp\left(\int_0^t c(X(s), V(s)\, ds\right) f(t, X(t), V(t)\right\}$$

is continuously differentiable for all $t > 0$,

$$\frac{d}{dt}\mathbf{E}_{x,v} \left\{ \exp\left(\int_0^t c(X(s), V(s))\, ds\right) f(t, X(t), V(t))\right\}$$

$$= \mathbf{E}_{x,v} \left\{ \exp\left(\int_0^t c(X(s), V(s))\, ds\right) \left(\frac{\partial f}{\partial t}(t, X(t), V(t))\right.\right.$$

$$\left.\left. + \mathcal{A}f(t, X(t), V(t)) + (cf)(t, X(t), V(t))\right)\right].$$

We deduce, by reasoning analogous to that of Theorem 2.3.4, the 'Feynman–Kac formula', which gives a probabilistic interpretation of the retrograde Kolmogorov equation:

Theorem 2.3.6 *Let $f, g \in C_b^1(\mathbf{R}^d \times E)$. Then*

$$u(t, x, v) = \mathbf{E}_{x,v} \left\{ \exp\left(\int_0^t c(X(\zeta), V(\zeta))\, d\zeta\right) g(X(t), V(t))\right.$$

$$\left. + \int_0^t \exp\left(\int_0^s c(X(\zeta), V(\zeta))\, d\zeta\right) f(X(s), V(s))\, ds\right\}$$

is the unique solution, in $C_b^1(\mathbf{R}_+ \times \mathbf{R}^d \times E)$, of the equation:

$$\begin{cases} \dfrac{\partial u}{\partial t}(t, x, v) = (\mathcal{A}u)(t, x, v) + c(x, v)u(t, x, v) + f(x, v), & t > 0, \ x \in \mathbf{R}^d, v \in E, \\ u(0, x, v) = g(x, v), & x \in \mathbf{R}^d, \ v \in E. \end{cases}$$

We shall now consider a generalization of the Fokker–Planck equation. Let $p \in C_b(\mathbf{R}^d \times E)$, $f \in C_b(\mathbf{R}_+ \times \mathbf{R}^d \times E)$, be such that

$$\sum_{v \in E} \int_{\mathbf{R}^d} p(x,v)\, dx = \alpha < \infty,$$

$$\sum_{v \in E} \int_{\mathbf{R}^d} f(t,x,v)\, dx = \beta(t) < \infty, \quad \forall t \geq 0,$$

where β is locally integrable over $\mathbf{R}^+$. We further assume that p and f are positive (the case p and f of arbitrary sign is treated in an analogous fashion by decomposing $p = p^+ - p^-$, $f = f^+ - f^-$). Set:

$$p(x,v) = \alpha \bar{p}(x,v), \quad f(t,x,v) = \beta(t)\bar{f}(t,x,v).$$

Theorem 2.3.7 *Let q be a mapping from $\mathbf{R}_+ \times \mathbf{R}^d \times E$ into $\mathbf{R}$, such that for all $t > 0$,*

$$\sum_{v \in E} \int_{\mathbf{R}^d} [|q|(t,x,v) + |\mathcal{A}^*q|(t,x,v)]\, dx < \infty,$$

satisfying the equation:

$$\frac{\partial q}{\partial t}(t,x,v) = (\mathcal{A}^*q)(t,x,v) + c(x,v)q(t,x,v) + f(t,x,v), \quad x \in \mathbf{R}^d, \ v \in E,$$

$$q(0,x,v) = p(x,v), \quad x \in \mathbf{R}^d, \ v \in E.$$

Then, for all $\phi \in C_b^1(\mathbf{R}^d \times E)$,

$$\sum_{v \in E} \int_{\mathbf{R}^d} q(t,x,v)\phi(x,v)\, dx$$

$$= \alpha \mathbf{E}_{0,\bar{p}}\left\{ \phi(X(t),V(t)) \exp\left(\int_0^t c(X(s),V(s))\, ds \right) \right\}$$

$$+ \int_0^t \beta(s)\mathbf{E}_{s,\bar{f}(s,\cdot)}\left\{ \phi(X(t),V(t)) \exp\left(\int_s^t c(X(r),V(r))dr \right) \right\} ds.$$

Proof Fix $t > 0$ and set for $0 \leq s \leq t$:

$$u(s,x,v) := \mathbf{E}_{s,x;v}\left\{ \phi(X(t),V(t)) \exp\left(\int_s^t c(X(r),V(r))\, dr \right) \right\}.$$

Then, u is the solution of the equation:

$$\frac{\partial u}{\partial s}(s,x,v) + (\mathcal{A}u)(s,x,v) + (cu)(x,v) = 0, \quad 0 \leq s \leq t,$$

$$u(t,x,v) = \phi(x,v).$$

Therefore, by using the equations satisfied by u and q,

$$\frac{d}{ds} \sum_{v \in E} \int_{\mathbf{R}^d} u(s,x,v)q(s,x,v)\,dx$$

$$= \sum_{v \in E} \int_{\mathbf{R}^d} \frac{\partial u}{\partial s}(s,x,v)q(s,x,v)\,dx + \sum_{v \in E} \int_{\mathbf{R}^d} u(s,x,v)\frac{\partial q}{\partial s}(s,x,v)\,dx$$

$$= \sum_{v \in E} \int_{\mathbf{R}^d} u(s,x,v)f(s,x,v)\,dx$$

$$= \beta \mathbf{E}_{s,\bar{f}(s,\cdot)} \left\{ \phi(X(t),V(t)) \exp \left(\int_s^t c(X(r),V(r))\,dr \right) \right\},$$

from which, by integrating from $s = 0$ to $s = t$,

$$\sum_{v \in E} \int_{\mathbf{R}^d} u(t,x,v)q(t,x,v)\,dx$$

$$= \sum_{v \in E} \int_{\mathbf{R}^d} u(0,x,v)q(0,x,v)\,dx$$

$$+ \int_0^t \beta(s)\mathbf{E}_{s,\bar{f}(s,\cdot)} \left\{ \phi(X(t),V(t)) \exp \left(\int_s^t c(X(r),V(r))\,dr \right) \right\}\,ds,$$

which is the stated formula. $\qquad\qquad\qquad\qquad\qquad\qquad\qquad\qquad$ □

2.4   Convergence to a diffusion

The aim of this section is to show how the first component of a transport process, suitably renormalized $\{(X_\epsilon(t), V_\epsilon(t))\}$, converges to a diffusion process as $\epsilon \to 0$. Likewise, the retrograde Kolmogorov equation associated with the random evolution converges to a 'heat equation', that is, a parabolic equation where the partial differential operator in x is second order. The results of this section will be stated without proof. We shall find, at the end of this chapter, some bibliographic comments concerning these results. It is not necessary to read this section to understand what follows, its aim is o make clear the remark made in Section 3.8 about the approximation of the solution of a transport equation as the free path tends to 0.

2.4.1   *The central limit theorem for a jump process*

We shall first state some results concerning the long-time behaviour of a jump Markov process.

We shall say that a jump Markov process with values in E, with infinitesimal generator $\{A(v,w); v,w \in E\}$, is *irreducible* if for all pairs

$v \neq w$ of states, there exists an integer n, and a chain of states $v_0 = v, v_1, \ldots, v_{n-1}, v_n = w$ such that $A(v_{i-1}, v_i) > 0$, $1 \le i \le n$.

Theorem 2.4.1 *Let $\{V(t), t \ge 0\}$ be an irreducible jump Markov process with values in E, with infinitesimal generator A. Then, there exists a probability Π over E which is the solution of the 'stationary Fokker–Planck equation':*

$$\Pi A \equiv 0, \qquad \text{that is} \qquad \sum_{w \in E} \Pi(w) A(w, v) = 0, \; v \in E.$$

If such a solution Π exists, then:

(a) *$\Pi Q_t = \Pi, t \ge 0$;*

(b) *for $y, v \in E$, $\lim_{t \to \infty} Q_t(y, v) = \Pi(v)$, and more generally for every probability μ over E, $(\mu Q_t)(v) \to \Pi(v), v \in E$;*

(c) *whatever the initial law μ_0:*

$$\frac{1}{t} \int_0^t \mathbf{1}_{\{V(s)=v\}} \, ds \to \Pi(v) \quad \text{almost surely}$$

as $t \to \infty$, $v \in E$, and more generally, for every function $f : E \to \mathbf{R}$, which is Π-integrable:

$$\frac{1}{t} \int_0^t f(V(s)) \, ds \to \sum_{v \in E} \Pi(v) f(v) \quad \text{almost surely}, \; t \to \infty.$$

The probability Π is called the *invariant probability* of the Markov process $\{V(t); t \ge 0\}$. Note that such a probability always exists in the case card $E < \infty$, which we assume for the rest of this section. (a) we say that if $\mu_0 = \Pi, \mu_t = \Pi$ for all $t > 0$. Moreover, if $\mu_0 = \Pi$, the process $\{V(t); t \ge 0\}$ is always *stationary* in the sense where for every $n \in \mathbb{N}$, $0 \le t_1 < t_2 < \cdots < t_n$, the law of the random vector $(V(t_1+s), V(t_2+s), \ldots, V(t_n+s))$ does not depend on $s \ge 0$ which follows from (a) and from Theorem 2.2.2. (b) we say that for all $\mu_0, \mu_t \to \Pi$ as $t \to \infty$. Finally, the result (c), called the ergodic theorem, is a generalization of the strong law of large numbers. It says that the proportion of the time spent between 0 and t in the state v converges to $\Pi(v)$ as t tends to infinity. We also have, in this context, a generalization of the central limit theorem.

Remark For the equation $f = Ag$, with $f \in L^2(E, \Pi)$ to have a solution $g \in L^2(E, \Pi)$, we must have $\langle \Pi, f \rangle = 0$ (in effect $\sum_v \Pi(v) f(v) = \sum_{v,v'} \Pi(v) A(v, v') g(v') = 0$).

We have the following result:

Theorem 2.4.2 *Suppose that the jump Markov process $\{V(t); t \geq 0\}$ is irreducible, and that it has an invariant probability Π. Let $f \in L^2(E, \Pi)$ satisfy $\langle \Pi, f \rangle = 0$, then there exists g such that $Ag = f$. We set:*

$$C(f) := -2 \sum_{v \in E} f(v)g(v)\Pi(v).$$

We have $C(f) > 0$ and further:

$$\frac{1}{\sqrt{tC(f)}} \int_0^t f(V(s))\, ds \to B,$$

in law as $t \to \infty$, where B is a reduced centred Gaussian random variable.

Theorem 2.4.3 *Moreover, the process:*

$$\left\{ \frac{1}{\sqrt{uC(f)}} \int_0^{tu} f(V(s))\, ds, \quad t \geq 0 \right\},$$

converges in law to a Brownian motion $\{B(t), t \geq 0\}$, as $u \to \infty$, that is, to a continuous centred Gaussian process such that $E[B(t)B(s)] = \inf(t, s)$.

Brownian motion will be reintroduced in Chapter 5.

Remark Under the hypotheses of Theorem 2.4.2, we can show that g can be written as a function of f in the form:

$$g(v) = -\int_0^\infty \mathbf{E}_v[f(V(t))]\, dt, \quad v \in E.$$

This results in the following formula for $C(f)$:

$$C(f) = 2 \int_0^\infty \mathbf{E}_\Pi[f(V(0))f(V(t))]\, dt,$$

where $\{V(t), t \geq 0\}$ is a stationary process with initial law Π, which (at least in the case where the Markov process is 'reversible') can be written as

$$C(f) = \int_{-\infty}^{+\infty} \mathbf{E}_\Pi[f(V(0))f(V(t))]\, dt,$$

subject to defining the stationary process $\{V(t)\}$ for all $t \in \mathbf{R}$.

Remark Theorem 2.4.2 can be extended to the case of a vector valued function b. Let $b \in L^2(E, \Pi; \mathbf{R}^d)$, such that $\langle \Pi, b_i \rangle = 0, 1 \le i \le d$. Let $C(b)$ be the $d \times d$ matrix such that:

$$(C(b))_{ij} = -\sum_{v \in E} (b_i(v)g_j(v) + b_j(v)g_i(v))\Pi(v), \quad \text{where } Ag_i = b_i.$$

Then, the matrix $(C(b))_{ij}$ is symmetric positive definite, we can define its square root $(C(b))^{1/2}$ and we have:

$$\left\{ \frac{1}{\sqrt{u}} \int_0^{tu} b(V(s)) \, ds, t \ge 0 \right\} \quad \to \quad C(b)^{1/2} B(t)$$

as $u \to \infty$ where $\{B(t), t \ge 0\}$ is a Brownian motion with values in $\mathbf{R}^d$, that is, a centred Gaussian process such that $\mathbf{E}[B(t)B(s)^*] = \inf(t, s)I$, where I denotes the identity matrix.

2.4.2 Convergence of a random evolution to a diffusion

Let $\{V(t), t \ge 0\}$ be an irreducible jump Markov process, which has a unique invariant probability Π. For every $\epsilon > 0$, we set:

$$V_\epsilon(t) = V(t/\epsilon^2), \quad t \ge 0,$$

and let $\{X_\epsilon(t), t \ge 0\}$ be the solution of the differential equation:

$$\begin{cases} \dfrac{dX_\epsilon(t)}{dt} = a(X_\epsilon(t)) + \dfrac{1}{\epsilon} b(V_\epsilon(t)), \\ X_\epsilon(0) = x_0. \end{cases}$$

We assume that $a : \mathbf{R}^d \to \mathbf{R}^d$, $b : E \to \mathbf{R}^d$ satisfies the hypotheses:

  (i) a is Lipschitz,
  (ii) $\langle \Pi, |b| \rangle < +\infty$, $\langle \Pi, b_i \rangle = 0$, $i = 1, \dots, d$.

It then follows from Theorem 2.4.2 that

$$\left\{ \frac{1}{\epsilon} \int_0^t b(V(\tfrac{s}{\epsilon^2})) \, ds, t \ge 0 \right\}$$

converges in law, as $\epsilon \to 0$, to $\{(C(b))^{1/2}B(t), t \ge 0\}$.

We then deduce, with the help of classical theorems on ordinary differential equations:

Theorem 2.4.4 *the process* $\{X_\epsilon(t), t \ge 0\}$ *converges in law, as* $\epsilon \to 0$, *to the solution* $\{X(t), t \ge 0\}$ *of the stochastic differential equation:*

$$\begin{cases} \dfrac{dX}{dt}(t) = a(X(t)) + (C(b))^{1/2} \dfrac{dB(t)}{dt}, \\ X(0) = x_0. \end{cases}$$

Note that as $B(t)$ is not differentiable in the usual sense (but only in the sense of distributions), the stochastic differential equation above is usually written in the following 'differential form':

$$\begin{cases} dX(t) = a(X(t))\, dt + (C(b))^{1/2} dB(t), & t \geq 0; \\ X(0) = x_0. \end{cases}$$

This notation is a convention, which means that

$$X(t) = x_0 + \int_0^t a(X(s))\, ds + (C(b))^{1/2} B(t), \quad t \geq 0.$$

We shall now give a generalization of Theorem 2.4.4.

$\{V_\epsilon(t); t \geq 0\}_{\epsilon > 0}$ being defined as above, we consider the ordinary differential equation perturbed by $\{V_\epsilon(t)\}$:

$$\begin{cases} \dfrac{dX_\epsilon(t)}{dt} = a(X_\epsilon(t), V_\epsilon(t)) + \dfrac{1}{\epsilon} b(X_\epsilon(t), V_\epsilon(t)), \\ X_\epsilon(0) = x_0, \end{cases}$$

and make the following hypotheses on the coefficients:

(i) For every $v \in E$, $x \to a(x, v)$ and $x \to b(x, v)$, which are Lipschitz mappings, from $\mathbf{R}^d$ into itself.

(ii) There exists $c > 0$ such that

$$\sum_v (|a(x, v)| + |b(x, v)|) \Pi(v) \leq c(1 + |x|)$$

(iii)

$$\sum_v b(x, v) \Pi(v) = 0, x \in \mathbf{R}^d,$$

and for all $1 \leq i \leq d$, there exists $g_i(x, v)$ such that

$$(Ag_i)(x, v) = b_i(x, v), \quad 1 \leq i \leq d, \quad x \in \mathbf{R}^d, \quad v \in E,$$

$$\sum_v (|b(x, v)|^2 + |g(x, v)|^2) \Pi(v) \leq c(1 + |x|^2).$$

Moreover, we define for all i, $1 \leq i \leq d$ and $x \in \mathbf{R}^d$:

$$\bar{a}_i(x) = \sum_v a_i(x, v) \Pi(v) - \sum_v b(x, v) \cdot \nabla_x g_i(x, v) \Pi(v),$$

$$C(b(x))_{ij} = -\sum_v [b_i(x, v) g_j(x, v) + g_i(x, v) b_j(x, v)] \Pi(v).$$

As before, the matrix C_{ij} is positive semidefinite and we can therefore define its square root. We assume that:

(iv) $\bar{a}_i$, $[C(b)]_{i,j}^{1/2} \in C^1(\mathbf{R}^d)$, $1 \leq i,j \leq d$, and has bounded first-order partial derivatives.

We then have, under the hypotheses above, the following convergence result:

Theorem 2.4.5 *The process $\{X_\epsilon(t); t \geq 0\}$ converges in law as $\epsilon \to 0$ to the solution $\{X(t); t \geq 0\}$ of the stochastic differential equation:*

$$\begin{cases} dX(t) = \bar{a}(X(t))dt + [C(b)]^{1/2}(X(t))dB(t), & t \geq 0, \\ X(0) = x_0. \end{cases}$$

We refer to Chapter 5 for a description of stochastic differential equations.

2.4.3 Convergence of the associated Kolmogorov equations

In the light of the Feynman–Kac contained in Theorems 2.3.6 and 2.3.7, the convergence theorem above implies the following convergence result on partial differential equations.

Let $u_\epsilon = u_\epsilon(t,x,v)$ be the unique solution (in the sense of Theorem 2.3.6) of the equation:

$$\begin{cases} \dfrac{\partial u_\epsilon}{\partial t}(t,x,v) = \left(a + \dfrac{1}{\epsilon}b\right) \cdot \nabla_x u_\epsilon + \dfrac{1}{\epsilon^2} \displaystyle\sum_{v' \in E} A(v,v')u_\epsilon(x,v') + d \cdot u_\epsilon \\ \qquad\qquad + f(x,v), \quad t \geq 0, \ x \in \mathbf{R}^d, \ v \in E, \\ u_\epsilon(0,x,v) = u_0(x,v), \quad x \in \mathbf{R}^d, \ v \in E. \end{cases}$$

Then, as $\epsilon \to 0$, we have:

$$u_\epsilon(t,x,v) \to u(t,x) \quad \forall (t,x,v) \in \mathbf{R}_+ \times \mathbf{R}^d \times E,$$

where $u(t,x)$ is the unique solution of the parabolic equation:

$$\begin{cases} \dfrac{\partial u}{\partial t}(t,x) = Lu(t,x) + \bar{d}(x)u + \bar{f}(x), \quad t \geq 0, \ x \in \mathbf{R}^d, \\ u(0,x) = \bar{u}_0(x), \quad x \in \mathbf{R}^d, \end{cases}$$

with:

$$L = \frac{1}{2}\sum_{i,j=1}^{d} C(b)_{ij}\frac{\partial^2}{\partial x_i \partial x_j} + \sum_{i=1}^{d} \bar{a}_i(x)\frac{\partial}{\partial x_i},$$

$$\bar{d}(x) = \sum_{v \in E} d(x,v)\Pi(v), \quad \bar{f}(x) = \sum_{v \in E} f(x,v)\Pi(v),$$

$$\bar{u}_0(x) = \sum_{v \in E} u_0(x,v)\Pi(v).$$

We have an analogous convergence result for 'Fokker–Planck type' equations.

A particular case, important in applications to transport equations, is the very simple case where $E \subset \mathbf{Z}^d$, $a = 0$, $b(v) = v$ (we say that b is the natural injection from $\mathbf{Z}^d$ into $\mathbf{R}^d$). In this case,

$$X_\epsilon(t) = x_0 + \frac{1}{\epsilon} \int_0^t V(s/\epsilon^2)\, ds.$$

We denote by C the positive definite matrix:

$$C_{ij} = \int_0^\infty \mathbf{E}[V_i(0)V_j(t) + V_j(0)V_i(t)]\, dt.$$

It follows from Theorem 2.4.2 that $X_\epsilon \to X$, where $X(t) = x_0 + (C)^{1/2}B(t)$. Further, $\{B(t), t \geq 0\}$ is a Brownian motion of dimension d.

We then show that $u_\epsilon = u_\epsilon(t, x, v)$, the solution of the transport equation,

$$\begin{cases} \dfrac{\partial u_\epsilon}{\partial t} &= \dfrac{1}{\epsilon} v \cdot \nabla_x u_\epsilon + \dfrac{1}{\epsilon^2} \sum_{v' \in E} A(v, v') u_\epsilon(x, v') \\ &\quad + d u_\epsilon + f \quad \text{for } t \geq 0, \ x \in \mathbf{R}^d, v \in \mathbf{Z}^d, \\ u_\epsilon(t, x, v) &= u_0(x, v), \quad x \in \mathbf{R}^d, \ v \in \mathbf{Z}^d, \end{cases}$$

converges to u, the solution of the parabolic equation:

$$\begin{cases} \dfrac{\partial u}{\partial t} = \dfrac{1}{2} \sum_{i,j=1}^d C_{ij} \dfrac{\partial u}{\partial x_i \partial x_j}(x) + \bar{d}(x)u(x) + \bar{f}(x), \quad t \geq 0, \quad x \in \mathbf{R}^d, \\ u(0, x) = \bar{u}_0(x), \quad x \in \mathbf{R}^d. \end{cases}$$

2.5 The general transport process

We shall generalize the idea of Section 2.3. The jump process $\{V(t); t \geq 0\}$ will now take values in $\mathcal{V} = \mathbf{R}^k$. Moreover, we shall consider some 'general random evolutions', whose discontinuous component $\{V(t)\}$ will no longer be a Markov process. Everything that is said here is equally valid in the case where $\mathcal{V}$ is an open set of $\mathbf{R}^k$ (in this case and if there is a nonzero derivative term in the velocity, it is necessary to specify the boundary conditions on $\partial \mathcal{V}$, see Section 3.4).

2.5.1 Jump Markov process

We shall now describe a jump Markov process $\{V(t); t \geq 0\}$ with values in $\mathbf{R}^k$. The process $\{V(t); t \geq 0\}$ will again be of the form:

$$V(t) = \sum_{\{n \geq 0; T_n < \infty\}} \xi_n \mathbf{1}_{[T_n, T_{n+1}[}(t),$$

where the times of jumps take their values in $\mathbf{R}_+ \cup \{+\infty\}$, $T_n < T_{n+1}$ on the set $\{T_n < \infty\}$, $T_n \to \infty$ almost surely as $n \to \infty$, and the random variables ξ_n take their values in $\mathbf{R}^k$.

We are given a transition kernel Π over $\mathbf{R}^k$, that is, $\forall v \in \mathbf{R}^k$, $\Pi(v,.)$ is a probability over $\mathbf{R}^k$ satisfying

(i) $\Pi(v, \{v\}) = 0$,
(ii) $\forall B \in \mathcal{B}_k$, $v \to \Pi(v, B)$ is a measurable mapping.

We assume that λ is a measurable mapping from $\mathbf{R}^k$ into $\mathbf{R}_+$, which satisfies:

$$\bar{\lambda} := \sup_v \lambda(v) < \infty.$$

First, we specify the conditional law of the pair (T_1, ξ_1), given that $V(0) = v$. Conditionally in $V(0) = v$, T_1 and ξ_1 are independent, T_1 follows an exponential law with parameter $\lambda(v)$ and the law of ξ_1 is given by $\Pi(v,.)$.

There are two ways to construct the process $\{V(t)\}$, which will be associated with two Monte-Carlo techniques (real shock or fictitious shock). We shall first describe the direct method (real shock): for all $n \geq 1$, the conditional law of $(T_{n+1} - T_n, \xi_{n+1})$ given (T_n, ξ_n) is the law:

$$\lambda(\xi_n)e^{-\lambda(\xi_n)t}\, dt\Pi(\xi_n, dv).$$

What we have done above completely specifies the conditional law of the infinite sequence $\{(T_n, \xi_n), n \geq 1\}$ given ξ_0, and therefore also the conditional law of $\{V(t), t > 0\}$ given $V(0)$.

We can also construct the process $\{V(t)\}$ by the following method of fictitious shocks: let $\{N(t), t \geq 0\}$ be a Poisson process with intensity $\bar{\lambda}$ whose times of jumps are denoted $S_1, S_2, \ldots$, and $\{\eta_n, n \in \mathbb{N}\}$ a Markov chain with values in $\mathbf{R}^k$ independent of $\{N(t)\}$, with transition kernel $\bar{\Pi}(v, dv')$ defined as follows:

$$\bar{\Pi}(v, dv') = \bar{\lambda}^{-1}[(\bar{\lambda} - \lambda(v))\delta_v(dv') + \lambda(v)\Pi(v, dv')],$$

where δ_v denotes the Dirac measure at the point v. We can then verify that the process:

$$V(t) := \sum_{n=0}^{\infty} \eta_n \mathbf{1}_{[S_n, S_{n+1}[}(t), t \geq 0$$

satisfies the properties stated above, as long as the law of η_0 coincides with that of ξ_0.

We then prove:

Theorem 2.5.1 $\{V(t), t > 0\}$ *is a homogeneous Markov process.*

Denoting by A the infinitesimal generator of $\{V(t); t \geq 0\}$, we have for $f \in C_b(\mathbf{R}^k)$:

$$(Af)(v) = \lambda(v)\left\{\int_{\mathbf{R}^k} f(v')\Pi(v, dv') - f(v)\right\}$$

$$= \lambda(v)\int_{\mathbf{R}^k} [f(v') - f(v)]\Pi(v, dv').$$

2.5.2   Transport processes and associated Kolmogorov equations

A first possible generalization of the 'elementary random evolutions' will consist of replacing the Markov process V with values in E by a jump Markov process with values in $\mathbf{R}^k$, as defined above. However, this generalization is not really satisfactory for our purposes. In the applications that we have in mind, $X(t)$ will represent the position of a particle, and $V(t)$ its velocity. In particular models, the velocity is assumed constant between shocks (which we call jumps). In others, it is not. But above all, there is no reason to assume that the law of the times of the shocks only depends on their velocity, and not on their position!

We shall now construct a more general class of random evolutions. We shall consider the process $\{Z(t) = (X(t), V(t)); t \geq 0\}$ with values in $\mathbf{R}^l$ ($l = d + k$). Initially, in order to simplify the description, we shall not distinguish between the coordinates $X(t)$ and $V(t)$ of $Z(t)$. We shall make clear later that only the velocity (and not the position) has a discontinuity at the time of a shock. Likewise, we shall write $z = (x, v)$.

To define the times of the successive jumps, we take a measurable bounded mapping λ from $\mathbf{R}^l$ into $\mathbf{R}_+$, denoted:

$$\bar{\lambda} = \mathrm{Sup}\,\lambda.$$

We also take a transition kernel Π over $\mathbf{R}^l$ which satisfies:

- $\forall z \in \mathbf{R}^l, B \to \Pi(z; B)$ is a probability over $\mathbf{R}^l$ such that $\Pi(z; \{z\}) = 0$.
- For every $B \in \mathcal{B}_l$, $z \to \Pi(z; B)$ is a Borel mapping.

Finally, we take $b \in C_b^1(\mathbf{R}^l; \mathbf{R}^l)$, denoting by b_1, b_2, respectively, the first d components of b and the k last. We now construct the process $Z(t)$:

- Between the jumps, this process satisfies the ordinary differential equation:

$$\frac{dZ(t)}{dt} = b(Z(t)).$$

- These jumps may be defined in two different ways:
  - Given a Poisson process $\{N(t), t \geq 0\}$ with intensity $\bar{\lambda}$, at the time of each jump t of this Poisson process, we take a random Bernoulli trial (independent of the Poisson process and everything else), to decide with probability $\bar{\lambda}^{-1}(\bar{\lambda} - \lambda(Z(t-)))$ to not modify Z, and with probability $\bar{\lambda}^{-1}\lambda(Z(t-))$ to 'make the jump' Z. In the case where Z jumps at time t, it does so independently of the Poisson process, of the Bernoulli trials, and of the preceding jumps, following the probability law $\Pi(Z(t-); dz)$. (We call this technique the method of 'fictitious shock'.)
  - Given a Poisson process $\{N(t), t \geq 0\}$ with intensity 1, let $0 < T_1 < T_2 < \cdots < T_n < \cdots$ be the times of successive jumps.

The nth jump of the process $\{Z(t), t \geq 0\}$ takes place at time S_n such that $\int_0^{S_n} \lambda(Z_s)\, ds = T_n$, $n \geq 1$, where $\{Z(t), t \geq 0\}$ is defined as the solution of the ordinary differential equation above, over each interval $[S_{n-1}, S_n)$, and the law of the jump at each time S_n is described as in the first stage. We remark that the probability that a jump takes place in the interval $[t, t + dt)$ is equal to $\lambda(Z(t))\, dt$, close to $(dt)^2$. (This is the 'real shock' technique.)

Denote by $\{\phi_z(t); t \geq 0\}$ the solution of the ordinary differential equation:

$$\frac{dZ(t)}{dt} = b(Z(t)), \quad t > 0; \quad Z(0) = z.$$

Also denote by $T_1, T_2, \ldots, T_n$, the times of jumps. We can then show that conditionally in $Z(0) = z$, T_1 and $Z(T_1)$ are independent, the conditional law of the time of the first jump T_1 is given by

$$\mathbf{P}(T_1 > t/Z(0) = z) = \exp\left\{-\int_0^t \lambda(\phi_z(s))\, ds\right\}, \quad t > 0,$$

and the conditional law of $Z(T_1)$, given that $Z(0) = z$ and $T_1 = t$, is given by

$$\mathbf{P}(Z(T_1) \in B/Z(0) = z, T_1 = t) = \Pi(\phi_z(t); B), \quad B \in \mathcal{B}_l.$$

We have therefore specified the conditional probability law $\Lambda_z(dt, dz')$ of the pair $T_1, Z(T_1)$, given that $Z(0) = z$, which can be stated explicitly by the formula:

$$\Lambda_z(dt, dz') = \lambda(\phi_z(t)) \exp[-\int_0^t \lambda(\phi_z(s))\, ds]\Pi(\phi_z(t); dz')\, dt.$$

More generally, for all $n \in \mathbb{N}$, the conditional law of $(T_{n+1} - T_n, Z(T_{n+1}))$ given $(T_n, Z(T_n))$ is the law $\Lambda_{Z(T_n)}(dt, dz)$. We can then show:

Theorem 2.5.2 *The process $\{Z(t); t \geq 0\}$ is a homogeneous Markov process.*

The infinitesimal generator of $\{Z(t)\}$ is the operator $\mathcal{A}$ which acts on the functions $\phi \in C_b^1(\mathbf{R}^l)$ as follows:

$$\mathcal{A}\phi(z) = b(z) \cdot \frac{\partial}{\partial z}\phi(z) + \lambda(z)\int_{\mathbf{R}^l} [\phi(z') - \phi(z)]\Pi(z; dz').$$

Having considered the associated Kolmogorov equations, we shall specify the form of the kernel Π. We assume that the paths of $\{X(t)\}$ are continuous, that is

$$\Pi(x, v; \{x\} \times \mathbf{R}^k) = 1.$$

We can therefore restrict ourselves to considering $\Pi(z; dz')$ as a measure over $\mathbf{R}^k$, which we shall denote $\Pi(x, v; dv')$. We shall assume from now on that the transition kernel is absolutely continuous with respect to the Lebesgue measure (this is not necessary for the retrograde Kolmogorov equation but it is necessary for the Fokker–Planck equation), that is that there exists a measurable mapping π from $\mathbf{R}^d \times \mathbf{R}^k \times \mathbf{R}^k$ into $\mathbf{R}_+$ such that

$$\Pi(x, v; dv') = \pi(x, v; v') dv'.$$

The operator $\mathcal{A}$ is written:

$$\mathcal{A}\phi(x, v) = b_1(x, v) \cdot \frac{\partial}{\partial x} \phi(x, v) + b_2(x, v) \cdot \frac{\partial}{\partial v} \phi(x, v)$$

$$+ \lambda(x, v)[\int_{\mathbf{R}^l} \phi(x, v') \pi(x, v; v') \, dv' - \phi(x, v)]$$

We have the 'Feynman–Kac formula' (see Theorem 2.3.6).

Theorem 2.5.3 *Let* c, f, $g \in C_b^1(\mathbf{R}^l)$. *If* λ *and* $\Pi(\cdot; dv')$ *are elements of* $C_b^1(\mathbf{R}^l)$, *then:*

$$u(t, z) =$$

$$\mathbf{E}_z \left\{ \exp\left(\int_0^t c(Z(\zeta)) \, d\zeta \right) g(Z(t)) + \int_0^t \exp\left(\int_0^s c(Z(\zeta)) \, d\zeta \right) f(Z(s)) \, ds \right\}$$

is the unique element of $C_b^1(\mathbf{R}_+ \times \mathbf{R}^l)$ *which is the solution of the equation:*

$$\begin{cases} \dfrac{\partial u}{\partial t}(t, z) = (\mathcal{A}u)(t, z) + c(z)u(t, z) + f(z), \quad z \in \mathbf{R}^l, \\[2mm] u(0, z) = g(z), \quad z \in \mathbf{R}^l. \end{cases}$$

We can also consider the 'Fokker–Planck' equations associated with the process $\{Z(t)\}$. We denote by $\mathcal{A}^*$ the adjoint operator of $\mathcal{A}$, that is,

$$\mathcal{A}^* \phi(z) = - \operatorname{div}_z(b\phi)(z)$$

$$+ \int_{\mathbf{R}^k} \lambda(x, v')\phi(x, v')\pi(x, v'; v) \, dv' - \lambda(x, v)\phi(x, v),$$

$$= \operatorname{div}_x(b_1(x, v)\phi(x, v)) + \operatorname{div}_v(b_2(x, v)\phi(x, v))$$

$$+ \int_{\mathbf{R}^k} \lambda(x, v')\phi(x, v')\pi(x, v'; v) dv' - \lambda(x, v)\phi(x, v).$$

We then know that if the law of $Z(0)$ is absolutely continuous with respect to the Lebesgue measure over $\mathbf{R}^l$, then so is the law of $Z(t)$. Assume that the density p_0 of $Z(0)$ is in $C_b^1(\mathbf{R}^l)$, then the density $p(t, z)$ of $Z(t)$ satisfies the Fokker–Planck equation:

$$\frac{\partial p}{\partial t}(z) = \mathcal{A}^* p(z),$$

$$p(0, z) = p_0(z).$$

Finally, we have the following generalization of the Fokker–Planck equation (see Theorem 2.3.7). Let c be a function in $C_b^1(\mathbf{R}^l)$. Let $p \in C_b^1(\mathbf{R}^l)$, $f \in C_b^1(R_+ \times \mathbf{R}^l)$ be positive, such that $\int_{\mathbf{R}^l} p(z)\,dz = \alpha < \infty$, $\int_{\mathbf{R}^l} f(t,z)\,dz = \beta(t) < \infty$, $t \geq 0$. We set, as in Section 2.3:

$$p(z) = \alpha \bar{p}(z), \quad f(t,z) = \beta(t)\bar{f}(t,z).$$

Theorem 2.5.4 *Let $q \in L^1((0,T) \times \mathbf{R}^l)$ (such that $\mathcal{A}^*q \in L^1((0,T) \times \mathbf{R}^l)$) be the solution of the equation:*

$$\begin{cases} \dfrac{\partial q}{\partial t}(z) = (\mathcal{A}^*q)(z) + c(z)q(z) + f(z), & t > 0, \ z \in \mathbf{R}^l, \\[2mm] q(0,z) = p(z), & z \in \mathbf{R}^l, \end{cases}$$

then for all $\phi \in C_b^1(\mathbf{R}^l)$,

$$\int_{\mathbf{R}^l} q(t,z)\phi(z)\,dz = \alpha \mathbf{E}_{0,\bar{p}}\left\{ \phi(Z(t)) \exp\left(\int_0^t c(Z(\zeta))\,d\zeta \right) \right\}$$
$$+ \int_0^t \beta(s)\mathbf{E}_{s,\bar{f}(s,\cdot)}\left\{ \phi(Z(t)) \exp\left(\int_s^t c(Z(\zeta))\,d\zeta \right) \right\} ds.$$

2.6   Application to transport equations

We now consider the case where $k = d$ and where $v \in \mathcal{V}$, which is a Borel set of $\mathbf{R}^d$. We consider the following equation satisfied by $u(t,x,v)$:

$$\begin{cases} \dfrac{\partial u}{\partial t} + v.\dfrac{\partial u}{\partial x} + \tau u = \mathcal{L}u + f, \\[2mm] u(0,\cdot) = g, \end{cases} \tag{2.3}$$

where $\tau = \tau(x,v)$ is a positive bounded function and

$$(\mathcal{L}\phi)(t,x,v) = \int_{\mathcal{V}} l(x,v,v')\phi(x,v')\,dv',$$

with l a positive bounded kernel. The functions f and g satisfy the same hypothesis as above.

We can generalize this model by considering a Vlasov equation (for this model, we must assume that $\mathcal{V}$ is an open set of $\mathbf{R}^d$):

$$\begin{cases} \dfrac{\partial u}{\partial t} + v \cdot \dfrac{\partial}{\partial x}u + \mathrm{div}_v(au) + \tau u = (\mathcal{L}u) + f, \\[2mm] u(0,\cdot) = g, \end{cases}$$

where $a : \mathbf{R}^d \times \mathcal{V} \twoheadrightarrow \mathbf{R}^d$, is such that $a_j(\cdot) \in C_b^1(\mathbf{R}^d \times \mathcal{V})$ and $a_j(x,\cdot)$ is in $C_b^1(\mathcal{V})$, for $0 \leq j \leq d$.

We can also consider the preceding equation or (2.3) over a spatial domain $\mathcal{D}$ which is an open set of $\mathbf{R}^d$. In this case, so that the problem is well posed, we have to add boundary conditions: $u(x, v)$ must be given for $x \in \partial\mathcal{D}$ and $v \in \mathcal{V}$ such that $\vec{n}_x \cdot \vec{v} > 0$ ($\vec{n}_x$ being the inward normal to $\partial\mathcal{D}$ in x). On the other hand, if $\mathcal{V}$ is not the whole of $\mathbf{R}^d$, we must take a condition on the boundary $\partial\mathcal{V}$ in a way that will be made precise in Chapter 4.

For the mathematical analysis of transport equations and their boundary conditions, we can consult, for example, Dautray and Lions, (1994).

From the numerical point of view, it is fundamental to distinguish two points of view, for problems (2.3) or (2.6) (and whatever the boundary conditions are).

A   We want to evaluate the solution u at a given point (or a finite number of such points).

B   We want to evaluate the solution u over the whole domain.

We remark that we must adopt point of view B if we have problems which are (weakly) nonlinear, for example, if the coefficients a, τ, l depend on the solution u.

If we adopt point of view A, it is then appropriate to consider the original equation (2.6) as a retrograde Kolmogorov equation (modified by the fact that the velocity of the particles is no longer v but $-v$ (see below)) and if we adopt point of view B, the original equation will be considered as a Fokker–Planck equation.

This dichotomy corresponds to two possible probabilistic interpretations of equation (2.6), according to whether we have considered it a 'retrograde Kolmogorov type' equation, or a 'Fokker–Planck type' equation.

Viewpoint A Interpretation as a retrograde Kolmogorov type equation.

We can put equation (2.6) into the form:

$$\frac{\partial u}{\partial t} = \mathcal{A}u + cu + f, \quad u(0) = g,$$

with

$$\mathcal{A}u = -v.\frac{\partial}{\partial x}u - a.\frac{\partial}{\partial v}u + \lambda(x, v) \int_{\mathcal{V}} (u(x, v') - u(x, v))\pi(x, v; v')\, dv'.$$

It is enough to choose:

$$\lambda(x, v) = \int_{\mathcal{V}} l(x, v; v')\, dv', \quad \pi(x, v; v') = \frac{l(x, v; v')}{\lambda(x, v)},$$

$$c(x, v) = \lambda(x, v) - \tau(x, v) - \sum_i \frac{\partial}{\partial v_i} a_i(x, v).$$

The mapping b introduced above then corresponds to

$$b(x, v) = \begin{pmatrix} -v \\ -a(x, v) \end{pmatrix}.$$

Viewpoint B   Interpretation as a Fokker–Planck type equation. We can put equation (2.6) in the form:

$$\frac{\partial u}{\partial t} = \mathcal{A}^* u + \tilde{c} u + f, \quad u(0) = g,$$

with

$$\mathcal{A}^* u = -v . \frac{\partial}{\partial x} u - \operatorname{div}_v (au)$$
$$+ \int_{\mathcal{V}} \tilde{\lambda}(x, v') u(x, v') \tilde{\pi}(x, v', v) \, dv' - \tilde{\lambda}(x, v) u(x, v).$$

It is enough to choose:

$$\tilde{\lambda}(x, v) = \int_V l(x, v', v) \, dv', \quad \tilde{\pi}(x, v', v) = \frac{l(x, v, v')}{\tilde{\lambda}(x, v)},$$

$$\tilde{c}(x, v) = \tilde{\lambda}(x, v) - \tau(x, v).$$

The mapping b introduced above then corresponds to

$$\tilde{b}(x, v) = \begin{pmatrix} v \\ a(x, v) \end{pmatrix}.$$

Similarly, these two interpretations are also useful the case of equation (2.3); in this case we must choose $a = 0$.

2.7   Bibliographic comments

There are many books that describe the jump Markov process with values in a finite or countable space E. We cite Bouleau (1988), Brémaud (1981), Cinlar (1975), and Karlin and Taylor (1981).

Sections 2.5 and 2.6 owe much to Dautray (1989). There is an almost complete presentation of 'random evolutions' in Pinsky (1991). The results of convergence to a diffusion are largely studied in Ethier and Kurtz (1986) and Kushner (1990). Breiman (1968) gives a presentation of Markov processes with values in general spaces. We can find most of the proofs of results stated in this chapter in Pardoux (1993). For the results of Section 2.4, we refer to Chapter 2 of Papanicolaou, (1975).

3

The Monte-Carlo method for the transport equations

In this chapter, we present the Monte-Carlo methods for the linear transport equations of the type described in Chapter 2. We shall first describe the principle of the two methods, known on the one hand as the 'adjoint Monte-Carlo method', which is based on the interpretation of the transport equation as a 'retrograde Kolmogorov type' equation (that is, we adopt viewpoint A of Section 2.6) and on the other hand 'direct Monte-Carlo method', which is based on the interpretation of the transport equation as a 'Fokker–Planck type' equation (viewpoint B). We shall then state convergence results for these methods, which follow easily from the results of Chapter 2, and from the law of large numbers.

This will be the subject of Sections 3.1 and 3.2, in the case where $x \in \mathcal{D} = \mathbf{R}^d$ and $v \in \mathcal{V} = \mathbf{R}^d$; the general case will be stated in Section 3.3, where we shall make clear how to take account of the boundary conditions.

We shall then detail, in Sections 3.4 and 3.5, the steps of the 'direct Monte-Carlo' method, when we use a discretization in time, as well as the techniques for evaluation of macroscopic quantities. Finally, in Section 3.6, we will study Monte-Carlo methods for stationary equations.

In Section 3.7, we shall show the limits of the method and the criteria for the choice of the time and space discretizations, which will allow us to address nonlinear problems. Finally, we shall quickly cover some specific techniques, notably the techniques of reduction of variance, in Sections 3.8 and 3.9.

We therefore consider the transport equation (or Vlasov equation) on the domain $\mathcal{D} \times \mathcal{V}$, given that $\mathcal{D}$ is an open set of $\mathbf{R}^d$ and that $\mathcal{V}$ is a Borel set of $\mathbf{R}^d$. We start by describing the principle of each of the two Monte-Carlo methods in the case where $\mathcal{V} = \mathbf{R}^d$ and $\mathcal{D} = \mathbf{R}^d$.

3.1 Principle of the adjoint Monte-Carlo method

While the direct method is well suited to finding the solution at all points on a grid, the adjoint method is better for finding $u(t, x, v)$ at a point (or at a small number of points). As we shall see below in the adjoint method, we generate the particles at the points where we are looking for the solution and we can also increase the precision of the solution by increasing the number of particles generated at only these points. We denote by $u(t, x, v)$

the solution of the transport equation that we have rewritten in the form of a retrograde Kolmogorov type equation (we adopt here viewpoint A):

$$\frac{\partial u}{\partial t} + v \cdot \frac{\partial u}{\partial x} + a \cdot \frac{\partial u}{\partial v} - cu - Hu = f, \tag{3.1}$$

$$u(0, \cdot) = g,$$

given that

$$Hu(x, v) = \lambda(x, v) \int_{\mathcal{V}} (u(x, v') - u(x, v)) \pi(x, v, v') \, dv'.$$

We must approximate, using Monte-Carlo, a quantity of the form:

$$u(t, x, v) = \mathbf{E}_{x,v} \left[g(X(t), V(t)) \exp \left(\int_0^t c(X(\zeta), V(\zeta)) \, d\zeta \right) \right.$$
$$\left. + \int_0^t f(s, X(s), V(s)) \exp \left(\int_0^s c(X(\zeta), V(\zeta)) \, d\zeta \right) \, ds \right].$$

For this, we generate N independent evaluations $\{X_i(s), V_i(s); 0 \le s \le t\}_{i=1,2,\ldots,N}$ of the random evolution $\{(X(s), V(s)\}$ (which is a Markov process). We do this in the following way. First, we set:

$$\begin{pmatrix} X_i(0) \\ V_i(0) \end{pmatrix} = \begin{pmatrix} x \\ v \end{pmatrix}.$$

• Between the jumps, (X_i, V_i) is the solution of the differential equation:

$$\frac{d}{ds} \begin{pmatrix} X_i(s) \\ V_i(s) \end{pmatrix} = - \begin{pmatrix} V_i(s) \\ a(X_i(s), V_i(s)) \end{pmatrix}.$$

• We simulate the jumps using one of the following two methods:
  ○ (1) The times of the jumps of the process $V_i(t)$ are those of the process $N(\int_0^t \lambda(X_i(s), V_i(s-)) \, ds)$ where $\{N(t)\}$ is a Poisson process of intensity 1. If this process has a discontinuity at time s, the conditional law of $V_i(s)$ given $V(s-)$ has density $\pi(X_i(s), V_i(s-), v) \, dv$.
  ○ (2) given $\bar{\lambda} = \max_{x,v} \lambda(x, v)$, we simulate a Poisson process with intensity $\bar{\lambda}$. Assume that this Poisson process has a discontinuity at time s (this time s is therefore an exponential random variable with parameter $\bar{\lambda}$). We then take a random sample independent of the Poisson process, of the preceding samples, and of the preceding jumps:
    ∗ with the probability $\bar{\lambda}^{-1}[\bar{\lambda} - \lambda(X_i(s), V_i(s-))]$ we do not modify $V_i(s)$;

* with the probability $\bar{\lambda}^{-1}\lambda(X_i(s), V_i(s-))$ we modify $V_i(s)$ by taking at random the new velocity $V_i(s)$ following the probability law:

$$\pi(X_i(s), V_i(s-), v)\, dv.$$

Method 2 is called the 'method of fictitious shock' and method 1 the 'method of real shock'. In method 2, the value $\bar{\lambda}$ can be taken as a local maximum.

• We then continue to solve the differential equation above until the next jump, and so on until time t. For each particle i, the calculation of the quantity:

$$M(t) = \exp\left(\int_0^t c(X(\zeta), V(\zeta))\, d\zeta\right)$$

reduces to solving the equation (by setting $w_i(t) = N^{-1}M_i(t)$):

$$\frac{dw_i(s)}{ds} = c(X_i(s), V_i(s))w_i(s), \quad w_i(0) = N^{-1}.$$

Then from the strong law of large numbers:

Theorem 3.1.1 *We have the following equality almost surely:*

$$u(t, x, v) = \lim_{N\to\infty} \sum_{n=1}^{N}\left[g(X_i(t), V_i(t))w_i(t) + \int_0^t f(s, X_i(s), V_i(s))w_i(s)\, ds \right].$$

It is clear that, thanks to this method, we can evaluate simply quantities of the following type:

$$I = \int_B \int_U u(x, v)\psi(x, v)\, dx\, dv, \quad \text{for } B \text{ a Borel set of } \mathcal{D}, U \text{ a Borel set of } \mathcal{V},$$

where ψ is an arbitrary integrable function. In effect, we can assume without loss of generality that ψ is positive with integral equal to 1; we then remark that

$$I = \mathbf{E}_\psi[g(X(t), V(t))\exp\left(\int_0^t c(X(\zeta), V(\zeta))\, d\zeta\right)$$

$$+ \int_0^t f(s, X(s), V(s))\exp\left(\int_0^s c(X(\zeta), V(\zeta))\, d\zeta\right)\, ds].$$

It is enough then to use the same algorithm as before, but taking initial conditions $(X_i(0), V_i(0))$ which are no longer deterministic, but chosen randomly from $B \times U$ according to the probability measure $\psi\, dx\, dv$.

3.2   Principle of the direct Monte-Carlo method

This method is better adapted than the one above for the calculation of
the quantity $u(t, x, v)$ for 'all the x, v'.

We depend on the interpretation of the transport equation as a 'Fokker–
Planck type' equation (we adopt viewpoint B) and we write it in the form:

$$\frac{\partial u}{\partial t} + v \cdot \frac{\partial u}{\partial x} + \text{div}_v(au) + ru - Ku = f, \qquad (3.2)$$

$$u(0, .) = g,$$

given that we have made the following changes in notation:

$$r = -\tilde{c},$$
$$\sigma = \tilde{\lambda},$$
$$k(x, v', v) = \tilde{\pi}(x, v', v).$$

With this notation, the operator K is written as

$$Ku(x, v) = \int_V \sigma(x, v')k(x, v', v)u(x, v') \, dv' - \sigma(x, v)u(x, v).$$

We can assume without loss of generality that f and g are positive. We
denote on the other hand

$$\alpha = \int \int g(x, v) \, dx \, dv, \quad \bar{p} = g/\alpha,$$

$$\beta = \int \int f(x, v) \, dx \, dv, \quad \overline{f} = f/\beta.$$

Now setting:

$$M(t) = \exp\left(-\int_0^t r(X(\zeta), V(\zeta)) \, d\zeta\right),$$

$$M^s(t) = \exp\left(-\int_s^t r(X(\zeta), V(\zeta)) \, d\zeta\right),$$

we then know that for every test function $\phi \in C_b(\mathcal{D} \times V)$, we have:

$$\int_{\mathcal{D}} \int_V u(t, x, v)\phi(x, v) \, dx \, dv$$

$$= \alpha \mathbf{E}_{0,\bar{p}}\left[\phi(X(t), V(t))M(t)\right] + \int_0^t \beta(s)\mathbf{E}_{s,\overline{f}(s,\cdot)}\left[\phi(X(t), V(t))M^s(t)\right] \, ds.$$

3.2.1   *Description of the method*

We shall first consider the case $f \equiv 0$. As before, we shall perform N evaluations of the Markov process $(X(t), V(t))$. The direct Monte-Carlo method consists of approximating the measure $u(t, x, v) \, dx \, dv$ by a linear combination of Dirac masses:

$$u(t, x, v) \, dx \, dv \simeq \alpha \sum_i w_i(t) \delta_{X_i(t)} \otimes \delta_{V_i(t)} \qquad (3.3)$$

(where δ denotes the Dirac measure $\delta_x(B) = 1$ if $x \in B$ and 0 otherwise) given that

$$w_i(t) = \exp\left(-\int_0^t r(X_i(\zeta), V_i(\zeta)) \, d\zeta\right) N^{-1}.$$

Each evaluation of the process $X_i(s), V_i(s), 0 \leq s \leq t, i = 1, 2, \ldots, N$ is generated in the following way:

- The N points $(X_1(0), V_1(0)), \ldots, (X_i(0), V_i(0)), \ldots$ are taken independently of one another, following the probability law $\bar{p}(x, v) \, dx \, dv$.
- The differential equation which follows each $(X_i(s), V_i(s))$ is this time

$$\frac{d}{ds}\begin{pmatrix} X_i(s) \\ V_i(s) \end{pmatrix} = \begin{pmatrix} V_i(s) \\ a(X_i(s), V_i(s)) \end{pmatrix}. \qquad (3.4)$$

  We note this process has a velocity opposite to what was used in the adjoint Monte-Carlo method.
- The process $(X_i(s), V_i(s))$ jumps in the same way as the adjoint method, given the function λ and the transition kernel π are replaced by the function σ and the kernel k (by using either the real shock technique or the fictitious shock technique).

We define the measure

$$\mu_{N,t}(dx, dv) = \alpha \sum_{i=1}^N w_i(t) \delta_{X_i(t)} \otimes \delta_{V_i(t)}.$$

From the law of large numbers, for every test function $\phi \in C_b(\mathcal{D} \times \mathcal{V})$,

$$\int_{\mathcal{D} \times \mathcal{V}} \phi(x, v) \mu_{N,t}(dx, dv) = \alpha \sum_{i=1}^N w_i(t) \phi(X_i(t), V_i(t))$$
$$\to \alpha \mathbf{E}_{0,\bar{p}}[M(t)\phi(X(t), V(t))],$$

almost surely as $N \to \infty$. In other words, we have:

Theorem 3.2.1 *The sequence of random measures $\{\mu_{N,t}\}$ converges almost surely strictly to the measure $u(t, x, v) \, dx \, dv$, as $N \to \infty$.*

We have taken the initial weights equal for all of the particles ($w_i(0) = N^{-1}$), but we obtain the same result if we take different weights satisfying, for a given constant C:

$$\sum_i w_i(0) = 1 \quad \text{and} \quad w_i(0) \leq CN^{-1}, \qquad \text{for all } i,$$

with the condition that the samples of $(X_i(0), V_i(0))$ are such that the measure:

$$\sum_{i=1}^{N} \delta_{X_i(0)} \times \delta_{V_i(0)}$$

converges in probability $\overline{p(x, v)}\, dx\, dv$ as N tends to infinity.

We shall now complete the description of the method in the case $f \neq 0$. We assume for simplicity that $f \geq 0$. This is a source term, which leads to the creation of new particles through time. We must approximate the quantity

$$\int_0^t \beta(s)\mathbf{E}_{s\bar{f}(s,\cdot)}[\phi(X(t), V(t))M^s(t)]\, ds.$$

We choose two sequences of integers $L(N)$—the number of time intervals—and $\ell(N)$—the number of particles—such that $L(N) \to \infty$, $\ell(N) \to \infty$ as $N \to \infty$.

We set $h(N) = t/L(N)$. We discretize the integral above, approximating it by the quantity:

$$\sum_{m=0}^{L(N)-1} h(N)\beta(mh(N))\mathbf{E}_{mh(N),\bar{f}(mh(N),\cdot)}[\phi(X(t), V(t))M^{mh(N)}(t)],$$

which we approximate by the following quantity:

$$\sum_{m=0}^{L(N)-1}\sum_{j=1}^{\ell(N)} \frac{h(N)}{\ell(N)}\beta(mh(N))\phi\left(X_j^{mh(N)}(t), V_j^{mh(N)}(t)\right) M_j^{mh(N)}(t),$$

where

- the $\ell(N)$ points $\{(X_j^{mh(N)}(mh(N)), V_j^{mh(N)}(mh(N))); 1 \leq j \leq \ell(N)\}$ are chosen independently of each other, following the density law $\bar{f}(ih(N), \cdot)$;

- the random functions $\{(X_j^{mh(N)}(s), V_j^{mh(N)}(s)), mh(N) \leq s \leq t\}$ evolve independently of one another, in the same way as the $(X_i(s), V_i(s))$;

- the $\{M_j^{mh(N)}(s), ih(N) \leq s \leq t, j = 1, 2, \ldots, \ell(N)\}$ are the solutions of the differential equations:

$$
\begin{cases}
\dfrac{dM_j^{mh(N)}(s)}{ds} \\
= -r(X_j^{mh(N)}(s), V_j^{mh(N)}(s))M_j^{mh(N)}(s), \quad mh(N) \leq s \leq t, \\
M_j^{mh(N)}(mh(N)) = 1.
\end{cases}
$$

Finally, in the general case $f \neq 0$, the measure $\mu_{N,t}$ takes the form:

$$
\mu_{N,t}(dx, dy)
$$

$$
= \frac{1}{N}\alpha \sum_{i=1}^{N} M_i(t)\delta_{X_i(t), V_i(t)}(dx, dy)
$$

$$
+ \frac{h(N)}{\ell(N)} \sum_{m=0}^{L(N)-1} \beta(mh(N)) \sum_{j=1}^{l(N)} M_j^{mh(N)}(t)\delta_{X_j^{mh(N)}(t), V_j^{mh(N)}(t)}(dx, dy),
$$

and Theorem 3.2.1 can be extended easily to this situation.

Here also, we have assumed that the initial weights are all equal for simplicity, but we can consider different weights as long as we approximate the density law $\overline{f(ih(N), \cdot)}$ well.

Remark The method described above can be seen as a description of the trajectories of physical particles whose evolution is modelled by the transport equation that we want to solve. This is why we speak of numerical particles or macroparticles.

Remark If r is positive and if σ and r are independent of the spatial variable x (or if we assume that they are locally independent of x), then the method of fictitious shocks reduces to the following procedure. Denoting by $\sigma_m(v) = \sigma(v) + r(v)$, the possible time of the jump t is simulated by a Poisson process with intensity σ_m. Then

- with probability σ/σ_m the velocity of the particle jumps,
- with probability r/σ_m the velocity of the particle is not modified.

In the two cases, the weight of particle i with velocity v_i, which was w_i^0 becomes:

$$
w_i(t) = w_i^0 \exp(-r(v_i)t)
$$

and the quantity $\mathbf{E}[\phi(X(t), V(t))M(t)]$ is the limit as N tends to infinity of:

$$
\sum_{i=1}^{N} w_i(t)\phi(X_i(t), V_i(t)).
$$

A variation of this method consists on the one hand of making a jump with probability 1, and on the other hand of modifying the weight of the particle at the time of each jump, by taking:

$$w_i(t) = w_i^0 \frac{\sigma(v_i)}{\sigma_m(v_i)}.$$

We then show that the calculation of $\mathbf{E}[\phi(X(t), V(t))M(t)]$ is not modified if we use this variation.

3.2.2   Link with particle methods

We have seen that the basic principle is to approximate the function u by a linear combination of Dirac masses in the space $\mathcal{D} \times \mathcal{V}$ following formula (3.3), then to follow the evolution of particles numerically according to the law corresponding to the infinitesimal generator of the semigroup of the equation. If there is no collision term, equation (3.2) becomes a simple advection equation and is called the Vlasov equation; and $\tau = r$ is interpreted as a damping coefficient. The corresponding semigroup is then very simple (the associated stochastic process is deterministic and corresponds to the trajectories defined by (3.4)).

We shall analyse what happens to the principle shown above when there is no jump in the velocity variable

1. We generate some initial particles to represent the initial data:

$$u(0, x, v) \, dx \, dv \simeq \sum_i w_i^0 \delta_{X_i^0} \otimes \delta_{V_i^0}. \tag{3.5}$$

2. We move each particle i following the equation of movement (3.4).

3. The weight w_i of each particle evolves in the following way as a function of time:

$$w_i(t) = w_i^0 \exp[-\int_0^t r(X_i(s), V_i(s)) \, ds]. \tag{3.6}$$

To take account of the source f, we generate some particles that evolve in the same way as above (except if the time of birth of each particle is different from 0). As we shall see later, we almost always discretize in time (this is indispensable in the case where we have to generalize the method to more complex equations). Thus, in each time step $[t^n, t^n + \Delta t]$, we must discretize the differential equation of movement (3.4) (different numerical schemes are possible depending on the physical context) and if necessary make an approximation of (3.6) for the calculation of $w_i(t^n + \Delta t)$ as a function of $w_i(t^n)$.

The scheme outlined is related to particle methods. These methods have been used for a long time in plasma physics, see, for example, Hocknoy and Eastwood, (1981) and Birdsall, (1991) (and the bibliography of Birdsall, 1991). We remark that the particle method itself is a method whose precision is arbitrarily large: the only discretization errors come from the initial errors (3.5) and the approximation of f, in the solution of the differential equations of movement and possibly the evaluation of (3.6). The only delicate point is passing from the particle representation (3.3) to a grid representation, which is in general necessary for the calculation of certain model parameters; we shall return to this subject below.

3.3   Boundary conditions

We consider the transport equation without an acceleration term (that is, in the case $a = 0$) over $\mathcal{D} \times \mathcal{V}$ where the domain $\mathcal{D}$ has a boundary denoted Γ:

$$\frac{\partial u}{\partial t} + v \cdot \frac{\partial u}{\partial x} + ru - Ku = f, \qquad (3.7)$$

$$u(0,.) = g.$$

We know that, for problem (3.7) to be well posed, we have to give a condition for each pair (x, v) such that x is on the boundary of Γ and such that

$$v \in Z_x,$$

denoting by n_x the inward normal at x to Γ and

$$Z_x = \{v/n_x \cdot v \geq 0\}.$$

We refer, for example, to Dautray and Lions, (1994) for the mathematical analysis of transport equations with boundary conditions. We shall only consider the three cases which are almost always met in practice:

1. Absorbing condition:

$$u(t, x, v) = 0, \quad x \in \Gamma, \ v \in Z_x.$$

2. Reflection condition:

$$u(t, x, v) = u(t, x, v - 2n_x(n_x \cdot v)), \quad x \in \Gamma, \ v \in Z_x.$$

3. Imposed inward flux:

$$u(t, x, v) = h(x, v), \quad x \in \Gamma, \ v \in Z_x,$$

where h is a given positive function.

Absorbing condition. The probabilistic interpretation allows us to say that we must follow the particles X_i, V_i as before until the moment where the particles arrive at the boundary Γ where they are destroyed.

Reflection condition. We follow the particles X_i, V_i as before until the moment where the particles arrive at the boundary Γ. They are reflected from the boundary with velocity:

$$\bar{V}_i = V_i - 2n_x(n_x \cdot V_i).$$

Imposed inward flux. We assume that there exists C such that for all x, we have:

$$\int_{Z_x} h(x,v)n_x \cdot v \, dv \leq C.$$

Remark We assume that $r = 0$ and $f = 0$ and we integrate (3.7) over $A \times V$ taking A to be an arbitrary open subset of $\mathcal{D}$ (such that $\Gamma_A = \Gamma \cap \bar{A}$ is nonempty). We obtain, denoting by $d\gamma(x)$ the surface measure over Γ,

$$\frac{\partial}{\partial t} \int_A \int_V u(x,v) \, d\gamma(x) \, dv - \int_{\partial A - \Gamma} \int_V u(x,v)v \cdot n_x \, d\gamma(x) \, dv$$

$$= \int_{\Gamma_A} \int_{Z_x} h(x,v)v \cdot n_x \, d\gamma(x) \, dv - \int_{\Gamma_A} \int_{V - Z_x} u(x,v)|v \cdot n_x| \, d\gamma(x) \, dv. \quad (3.8)$$

The second integral on the right-hand side corresponds to a flux of physical particles leaving the domain, and the first integral of the right-hand side corresponds to a flux of physical particles entering the domain. For every open set A, we have an estimate of $\int_A \int_V u(x,v) \, dx \, dv$ thanks to

$$\sum_{i/X_i \in A} w_i.$$

To simulate the inward flux numerically we proceed in the following way. On each time interval δt and each part of the boundary $\delta \Gamma$ we generate particles with index i having characteristics (X_i, V_i, w_i) coming into $\mathcal{D}$. To be consistent with the formula above, the weights must satisfy:

$$\sum_{i/ \text{ entering particles}} w_i = \int_{Z_x} h(x,v)v \cdot n_x \, dv \cdot \gamma(\Gamma) \, \delta t.$$

That is, we generate particles with surface density $H(x)$ given by

$$H(x) = \int_{Z_x} h(x,v)v \cdot n_x \, dv.$$

The law of distribution of the velocity V of these particles is given by the following result:

Proposition 3.3.1 *For fixed x and for each particle i, the law of distribution of the velocity V_i over Z_x is given by*

$$\frac{1}{H(x)} h(x, v) v \cdot n_x \, dv.$$

Proof *(heuristic)* We remark first that this problem is independent of the collision operator K, we can therefore remove the operator. We return to the notation used above. We consider a part A of $\mathcal{D}$ such that its 'boundary' $\Gamma_A = \partial A \cap \Gamma$ is a small surface element (with measure $\gamma(\Gamma_A)$) and such that n_x may be considered constant for $x \in \Gamma_A$. For every part δV of Z_x, by integrating (3.7) with $r = 0$, $f = 0$, we obtain:

$$\frac{\partial}{\partial t} \int_A \int_{\delta V} u(t, x, v) \, dx \, dv - \int_{\partial A - \Gamma} \int_{\delta V} u(t, x, v) n_x \cdot v \, d\gamma(x) \, dv$$

$$= \int_{\Gamma - A} \int_{\delta V} h(x, v) n_x \cdot v \, d\gamma(x) \, dv.$$

We introduce the quantity:

$$Q_{\delta V}(t) = \int_A \int_{\delta V} u(t, x, v) \, dx \, dv,$$

which is interpreted as the number of particles in A having their velocity in δV. The variation during a small time interval δt of this quantity $Q_{\delta V}(t)$ due to the entrance of particles can be approximated by

$$\gamma(\Gamma_A) \delta t \int_{\delta V} h(x, v) n_x \cdot v \, dv,$$

from where we have the proposition. □

Specific case. If h only depends on v through $|v|$ and therefore satisfies:

$$h(x, v) = \bar{h}_x(|v|),$$

then by introducing the direction cosine of v with respect to the normal to the surface:

$$\mu = n_x \cdot v / |v|,$$

and the azimuthal angle ϕ, defined over $[0, 2\pi]$, the law of distribution of v is

$$\frac{1}{H(x)} \bar{h}_x(z) z^3 \, dz \cdot \mu \, d\mu \cdot d\phi, \quad \text{where } z = |v|.$$

This probability law is then known as *Lambert's law* (see, e.g. Spanier and Gelbard, 1969). We now consider the case of the Vlasov equation on a domain $\mathcal{D}$ whose boundary is Γ. We must then apply the condition described above over Γ. Further, if the domain $\mathcal{V}$ is an open set of $\mathbf{R}^d$ and if we have a condition of the type:

$$u(x, v) = 0 \quad \text{for all} \quad v \in \partial\mathcal{V} \quad \text{t.q.} \quad a(x, v).\zeta_v \geq 0$$

(where ζ_v is the inward normal to $\partial\mathcal{V}$) then it is enough to stop the process $V(t)$ when it reaches the boundary $\partial\mathcal{V}$.

3.4   General scheme with time discretization

We shall apply the principle described above by introducing a time discretization into finite intervals $[t^n, t^n + \Delta t]$. This time discretization is not indispensable for very simple problems where the coefficients of the equation only depend on the spatial variable (as above), but it is indispensable when the coefficients depend on time or are functions of solutions of other equations coupled with the transport equation. On the other hand, in the very simple case stated above, we use this discretization most of the time for simple reasons of data management.

All that follows can be applied, with the necessary changes, to the Vlasov equation, but for simplicity we only consider the case of simple transport where the acceleration term a is zero, that is equation (3.7).

To start, we generate some particles representing the initial data $u(0)$, using formula (3.5). At each time step $[t^n, t^n + \Delta t]$, we make use of the following steps (denoting $t^{n+1} = t^n + \Delta t$):

1. We generate particles to take account of the source f.
2. For each particle i, we introduce a lifetime θ, which we initialize by Δt.
3. We displace each particle i according to the equation of movement:

$$\frac{\partial X_i}{\partial t} = V_i(t) \tag{3.9}$$

   or (3.4) in the case of 'Vlasov'.

4. For each particle i the free path process is stopped at the time τ, defined as the moment of the first jump of the process (N being a Poisson process)

$$t \mapsto N\left(\int_0^t \sigma(X_i(t^{n+1} - \theta + s), V_i(t^{n+1} - \theta + s))\, ds\right).$$

If $\tau \leq \theta$, we keep the characteristics of the particle, except the speed of particle i which was $V_i(\tau_-)$ and becomes $V_i(\tau)$ a random variable distributed on $\mathcal{V}$ according to the probability:

$$k(X_i(\eta), V_i(\eta_-), v)dv.$$

After the jump, we continue to follow the particle as in stage (2) by replacing θ by $\theta - \tau$.

5. At the end of the time step $[t^n, t^{n+1}]$ the weight of each particle i which was w_i^n becomes

$$w_i^{n+1} = w_i^n \exp\left[-\int_{t^n}^{t^{n+1}} r(X(s), V(s))\, ds\right].\qquad(3.10)$$

An alternative to the method described above to determine the time of the jump is to use the fictitious shock method described before (and whose application will be detailed in Section 3.8). If we do not use the fictitious shock method, as the function σ depends in general on x and on v and can vary very strongly as a function of x, we see that point (3) above implies that we must have a space grid. Generally, we therefore consider a grid $\mathcal{D}$ such that over each element m, the coefficients r and σ are assumed only to depend on v and are written $r_m(v), \sigma_m(v)$. We shall proceed in the following way over each time step $[t^n, t^n + \Delta t]$ and for each particle i.

Algorithm The characteristics, at time t^n, of the particle i are

$$\begin{bmatrix} X_i^n, V_i^n \\ w_i^n \end{bmatrix}$$

1. We initialize an event counter $q = 0$ and the lifetime left to each particle i which is $\theta_0 = \Delta t$. We set:

$$w_0 = w_i^n, \quad X_0 = X_i^n, \quad V_0 = V_i^n,$$

m_0 the number of the element such that $X_i^n \in m_0$.

2. Calculation of the 'free path' trajectory (that is (3.9) in the case where a is zero), which allows us to evaluate the exit time t_q from the element m_q.

3. Calculation of the stopping time τ_q following an exponential law with parameter $\sigma_{m_q}(V_q)$. For this we choose a random number y uniformly distributed over $[0,1[$ and we set:

$$\tau_q = -\log y/\sigma_{m_q}(V_q).\qquad(3.11)$$

4. We set $s = \text{Min}(t_q, \tau_q, \theta_q)$, we advance the particle:

$$X_{q+1} = X_q + sV_q$$

and we reset the weights:

$$w_{q+1} = w_q \exp(-r_{m_q}(V_q)s)\qquad(3.12)$$

(a) if $s = \theta_q$ (that is, we are at the end of a time step) we store all the characteristics (position, X_{q+1}, velocity, V_{q+1} and weights, w_{q+1}) and we go on to the next particle;

(b) if $s = \tau_q$ we choose a new velocity V_{q+1} in $\mathcal{V}$ according to the law

$$k_{m_q}(V_q, w)\, dw,$$

we set:

$$\theta_{q+1} = \theta_q - s, \quad m_{q+1} = m_q,$$

and we return to point (b) incrementing the counter q;

(c) if $s = t_q$ we find ourselves in a new element denoted by m_{q+1} and we set:

$$\theta_{q+1} = \theta_q - s,$$

and we return to point (b) incrementing the counter q.

Remark If the acceleration a is nonzero (the case of the Vlasov equation), we must solve a more complex movement equation than (3.9):

$$\frac{\partial X}{\partial t} = V, \qquad \frac{\partial V}{\partial t} = a(X, V). \tag{3.13}$$

More precisely, we always assume that V_i is constant along the path in the interior of an element and between two jumps.

To take account of boundary conditions of the type:

$$u(t, x, v) \quad \text{given for} \quad v \in Z_x,$$

where Z_x is defined in Section 3.3, we must generate particles in the way that has been described in Chapter 2, given that every particle leaving the domain of calculation is abandoned. The scheme above is a model scheme that we must adapt to each specific case; in particular, the results will only be viable if we have a sufficient number of particles per element (it leads to verifying throughout the calculation that this number is greater than some constant which depends on the desired precision).

3.5   Evaluation of the mesh quantities

In most problems, we need a representation 'on the mesh' of the two first moments of u, the density ϕ, and the current J, sometimes for diagnostics, sometimes to be used in an equation coupled to the principal equation:

$$\phi(t, x) = \int u(t, x, v)\, d\vec{v}, \tag{3.14}$$

$$\vec{J}(t, x) = \int \vec{v} u(t, x, v)\, d\vec{v}. \tag{3.15}$$

Depending on what one wants to do with these quantities, different methods of evaluation are possible. We shall describe some for the simple transport equation (3.7) with $f = 0$. We have an approximation of $u(t, \cdot)$ in the form:

$$u(t, x, v)\, dx\, dv = \sum_p w_p(t)\delta_{X_p(t)}\delta_{V_p(t)}.$$

Estimation by indicated value. This is a matter of evaluating the densities ϕ_M^n in a consistent way, in the elements M (with volume V_M) at the time t^n and the currents $j_F^{n+1/2} = \vec{J}^{n+1/2} \cdot \vec{n}_F$ on a face F (with surface S_F and normal $\vec{n}_F$) in a time interval $I = [t^n, t^{n+1}[$. We then set:

$$\phi_M^n = \frac{1}{V_M} \sum_{p/X_p(t^n)\in M} w_p(t^n),$$

$$j_F^{n+1/2} = \frac{1}{\Delta t S_F} \sum_{\substack{p/\exists t_p \in I^n \\ X_p(t_p) \in F}} w_p(t_p)\, \mathrm{sign}(\vec{n}_F \cdot \vec{V}_p(t_p)).$$

We then verify that if $r = 0$, we have:

$$V_M(\phi_M^{n+1} - \phi_M^n) = \Delta t \sum_{F,\ \text{faces of } M} \pm j_F S_F. \qquad (3.16)$$

The sign $\pm$ depends on the orientation of the normal.

In the case where $r = r(x) \neq 0$, the difference between the two members of (3.16) corresponds to the loss of particles:

$$Q = \int_I \int_M \int_V r(x)u(t, x, v)\, dx\, dv\, dt.$$

This quantity can also be evaluated by the following expression:

$$Q_M^{n+1/2} = \sum_{p/X_p[I]\cap M\neq\emptyset} w_p(t_p^M)[1 - \exp(-r_M(\bar{t}_p^M - t_p^M))],$$

where t_p^M and $\bar{t}_p^M$ denote, respectively, the times of entrance and exit of the particle p into the element M.

This allows us to have another estimate ϕ during a time step given by the formula:

$$\phi_M^{n+1/2} = \frac{1}{\Delta t V_M r_M} Q_M^{n+1/2}.$$

Estimation by 'time of passage'. The remark above suggests a general estimate of ϕ valid for arbitrary r (independent of v, zero or not), in effect we can take:

$$\phi_M^{n+1/2} = \frac{1}{V_M} \left[\sum_{p/X_p \cap M \neq 0} (\bar{t}_p^M - t_p^M) \right]^{-1}$$

$$\times \sum_{p/X_p \cap M \neq 0} \frac{1}{2}(w_p(t_p^M) + w_p(t_p^{\bar{M}}))(\bar{t}_p^M - t_p^M).$$

In this formula, we see that

$$\sum_p (\bar{t}_p^M - t_p^M)$$

can be considered as a particle 'estimate' of the constant Δt.

On the other hand, we can verify that

$$r_M \sum_p \frac{1}{2}(w_p(t_p^M) + w(\bar{t}_p^M))(\bar{t}_p^M - t_p^M)$$

can be considered as an estimate of the loss of particles $Q_M^{n+1/2}/(\Delta t V_M)$, if we replace formula (3.10) by the following formula:

$$w_p(\bar{t}_p^M) = w_p(t_p^M)\frac{1 - r_M(\bar{t}_p^M - t_p^M)/2}{1 + r_M(\bar{t}_p^M - t_p^M)/2}$$

(which is a classical discretization of $\partial w/\partial t + rw = 0$ on the time interval $[t_p^M, \bar{t}_p^M]$).

Estimation by 'shape functions'. This is a matter of evaluating ϕ and $\vec{J}$ at the end of the time step and at the nodes of the grid. Denoting a node by A_i, we can associate, in a natural way, a linear (or bilinear) shape function X_i which is such that

$$X_i(A_i) = 1 \quad X_i(A_j) = 0 \quad j \neq i,$$

$$\sum_i X_i(x) = 1 \quad \forall x.$$

We can also use more complex shape functions only satisfying $\sum_i X_i(x) = 1$ for all x, whose support can be extended along several elements around A_i. We then evaluate ϕ and $\vec{J}$ in the following way:

$$\phi_i^n = \left[\sum_p X_i(X_p^n)w_p^n \right] \cdot \left[\sum_p X_i(X_p^n) \right]^{-1},$$

$$\vec{J}_i^n = \left[\sum_p X_i(X_p^n)w_p^n \vec{V}_p^n \right] \cdot \left[\sum_p X_i(X_p^n) \right]^{-1}.$$

Here, the quantity $\sum_p \mathcal{X}_i(X_p^n)$ is an 'estimate' of the unit of volume (see, e.g. Raviart, 1985; Birdsall, 1991).

3.6   Stationary problems

Frequently, we must solve the stationary transport equations, where we look for a function $u = u(x, v)$, which is the solution of the following equation:

$$v \cdot \frac{\partial u}{\partial x} + ru - Ku = f. \tag{3.17}$$

If $\mathcal{D}$ is not equal to R^d, we must impose some boundary conditions. To fix ideas, we assume that on one part Γ of $\partial\mathcal{D}$ we have a zero inward flux condition:

$$u(x, v) = 0 \quad \forall v \quad \text{t.q.} \quad v.n_x \geq 0 \quad (n_x, \text{ inward normal } \Gamma)$$

and that over $\partial\mathcal{D}\Gamma$ we have a reflection condition.

It is well known that the operator $\Psi \to v \cdot (\partial/\partial x)\Psi - K\Psi$ has a very small positive eigenvalue λ (if $\mathcal{D} = R^d$, then $\lambda = 0$; if $\mathcal{D}$ is bounded and if Γ has nonzero surface measure, then λ is strictly positive).

We make the hypothesis:

$$\inf(r(x, v)) > -\lambda. \tag{3.18}$$

Proposition 3.6.1 *For all f in $L^2(\mathcal{D} \times \mathcal{V})$, there exists a unique solution u of equation (3.17). Further, if $U(t)$ is the solution of:*

$$\frac{\partial U}{\partial t} + v \cdot \frac{\partial U}{\partial x} + rU - KU = 0, \quad U(0) = f$$

with the same boundary conditions as above, then $\|U(t)\|_{L^2}$ decreases exponentially fast as t tends to infinity and the solution u is given by the formula:

$$u(x, v) = \int_0^\infty U(t, x, v) \, dt.$$

Finally, for all $\phi \in_{C_b} (\mathcal{D} \times \mathcal{V}), f \geq 0$, we have:

$$\int \int u(x, v)\phi(x, v) \, dx \, dv$$

$$= \beta \mathbf{E}_{0,\overline{f}} \left[\int_0^\infty \exp\left(\int_0^t -r(X(s), V(s)) \, ds \right) \phi(X(t), V(t)) \, dt \right], \tag{3.19}$$

where $\beta = \int_{\mathcal{D} \times \mathcal{V}} f(x, v) \, dx \, dv$, $\overline{f} = f/\beta$.

Proof Let T_t be the semigroup of operators over $L^2(\mathcal{D} \times \mathcal{V})$ defined for arbitrary f by $T_t f = U(t)$, for U defined above. Denote by $\eta(t)$ the solution of

$$\frac{\partial \eta}{\partial t} + v \cdot \frac{\partial \eta}{\partial x} - K\eta = 0, \quad \eta(0) = f$$

with the same boundary conditions as above. We know, thanks to the properties of the transport operator (see, e.g. Dautray and Lions, 1994), that:

$$\|\eta(t)\|_{L^2} \le Ce^{-\lambda t}$$

and if $r_0 = \inf r(x, v)$ we verify easily that:

$$\|U(t)\|_{L^2} \le \|\eta(t)\|_{L^2} e^{-r_0 t} \le Ce^{-r_0 t - \lambda t}.$$

In this way, we can define $u = \int_0^\infty U(t, x, v)\, dt$. We know that for h tending to 0, we have for all g sufficiently regular in x:

$$\lim_h \frac{1}{h}(T_h g - g) = -\left(v \cdot \frac{\partial g}{\partial x} + rg - Kg\right)$$

from which we deduce:

$$\lim_h \frac{1}{h} \int_0^h T_t f\, dt = \lim_h \frac{1}{h}\left(\int_0^\infty T_t f\, dt - \int_0^\infty T_{t+h} f\, dt\right)$$

$$= -\lim_h \frac{1}{h}(T_h - T_0) \int_0^\infty T_t f\, dt = \left(v \cdot \frac{\partial}{\partial x} + r - K\right)\left(\int_0^\infty U(t, x, v)\, dt\right).$$

Since $f \to T_t f$ is continuous with respect to t, from which we deduce that u satisfies (3.17).

The last relation (3.19) is a direct consequence of the probabilistic interpretation of $U(t)$ given in Chapter 2 ('Viewpoint B, interpretation of the transport equation as a Fokker–Planck type equation'). □

3.6.1 General scheme

We have seen that if f is positive, then u will be positive. The principle of the method for the numerical solution of (3.17) is then the following in the case where f is positive and $\int f(x, v)\, dx\, dv = 1$.

We generate a number, N, of random variables (X_i^0, V_i^0), which are independently distributed following the probability density law $f(\cdot)$. Each random variable is the initial state of a particle i, we assign a weight to this particle w_i^0, the weights being such that

$$\sum_{i=1}^N w_i^0 = 1,$$

each particle moves (in $\mathcal{D} \times \mathcal{V}$) as the realization of the process $(X(t), V(t))$. We denote by $(X_i(t), V_i(t))$ the characteristics (position, velocity) of the particle i at the time t, and $w_i(t)$ its weight, which is the solution of the same ordinary differential equation as above (see the method for evolution equations).

Assume $\mathcal{D}$ is bounded. Even if it means slightly modifying $\mathcal{V}$, we can assume (when σ and k are bounded close to $v = 0$) that $\inf_{v \in \mathcal{V}} |v| > 0$, then we show that the exit time of each particle is almost surely finite. Thanks to Proposition 3.6.1, we obtain the result below, denoting by ψ^N the measure over $\mathcal{D} \times \mathcal{V}$ defined by

$$\psi^N = \sum_{i=1}^{N} \int_0^\infty w_i(t) \delta_{X_i(t)} \delta_{V_i(t)} \, dt.$$

Corollary 3.6.2 *As N tends to infinity, we have:*

$$\int \int \psi^N (\, dx \, dv) \phi(x, v) \to \int \int u(x, v) \phi(x, v) \, dx \, dv.$$

The practical implementation is made in the same way as for the evolution equations except that here we follow the particles until they leave the domain of calculation $\mathcal{D}$.

3.6.2 *Evaluation of the mesh quantities*

The evaluation of the mesh quantities is done following the same principle as above (with the notation above). For values of the current $\vec{J}$, we can use the estimator:

$$j_F = \frac{1}{S_F} \sum_{p/\exists t_p\ X_p(t_p) \in F} w_p(t_p) \ \text{sign}(\vec{n}_F \cdot \vec{V}_p(t_p)).$$

For the evaluation of the density ϕ_M, we cannot use the method of 'indicated value'; on the other hand, the estimator by 'time of passage' defined above gives good results.

For the density or the current, we can also use the formulae obtained thanks to the shape functions.

3.7 Limits of the method and generalizations

3.7.1 *Limits of the method*

We can treat problems of very large size without too much difficulty by adjusting the number of particles (and therefore the cost of the calculation) to the required precision. But we must be aware of the limits of the method that we shall now describe.

Denote by $\bar{v}$ a characteristic particle velocity, $\bar{\sigma}$ and $\bar{r}$ are characteristic values of the coefficients σ and r.

A priori, the spatial discretization depends on the physics of the problem; the time step can depend on the way in which the source S evolves and additionally for reasons of efficiency the time step Δt will be such that a particle with velocity $\bar{v}$ does not cross more than a fixed number of elements. That is, denoting by $\overline{\Delta x}$ an average grid step:

$$\bar{v}\Delta t \leq C\overline{\Delta x}, \quad \text{with } C \text{ of order 1.} \tag{3.20}$$

On the other hand, we remark that the algorithm described above will only be effective if

$$\frac{\overline{\sigma_m}^{-1}}{\Delta t} \quad \text{is not much less than 1}$$

or if

$$\frac{\bar{v}\overline{\sigma_m}^{-1}}{\overline{\Delta x}} \quad \text{is not much less than 1.}$$

That is, in each element m, the average free path should not be much smaller than $\overline{\Delta x}$, otherwise the particles will have to change velocity many times before leaving the element m and the calculation time will become prohibitive. The criterion above is explained well if we refer to the physics of the problem. If $\lambda = \bar{v}\bar{\sigma}^{-1}$ is much less than the characteristic dimensions (and further if r is much less than σ), we know that the solution of the transport equation is very close to the solution of a diffusion equation satisfied by $U(x)$:

$$\frac{\partial U}{\partial t} + \bar{r}U - \frac{\partial}{\partial x}\left(\frac{1}{3\bar{\sigma}}\frac{\partial U}{\partial x}\right) = \bar{S}.$$

See paragraph 4 of the preceding chapter (or from the perspective of partial differential equations, e.g. Dautray and Lions, 1994; Papanicolaou, 1975; Ringeissen and Sentis, 1991). We know that the discretization of such parabolic equations is not done easily with explicit time schemes.

However, in the case where λ is much smaller than the size of an element, we can think of replacing the many small jumps over $\mathcal{V}$ and the small free paths between the jumps by a unique jump in $\mathcal{D} \times \mathcal{V}$ whose characteristics are exactly calculated by solving a diffusion equation (see Giorla and Sentis, 1987, for a description of this method).

3.7.2 Operator splitting

A first criterion on the time step has just been shown in (3.20). But in certain situations (particularly if the coefficients of the transport equation are not constant in time or if the source term depends on the solution), it may be interesting to carry out a splitting between the 'advection' part and the 'collision' part at each time step, and this will induce another condition on the time step. Therefore, over a time step $[t^n, t^n + \Delta t]$ we solve successively:

$$\frac{\partial \phi}{\partial t} + v \cdot \frac{\partial \phi}{\partial x} + r\phi = f, \quad \phi(t^n) = \bar{u}^n,$$

$$\frac{\partial \psi}{\partial t} - K\psi = 0, \quad \psi(t^n) = \phi(t^n + \Delta t),$$

and we define

$$\bar{u}^{n+1} = \psi(t^n + \Delta t).$$

It is clear that with this scheme the numerical solution is simpler: in the first phase, we displace particles along the free path trajectory and we modify their weights using formula (3.10); in the second phase, we treat the collisions by choosing successive stopping times τ_q (using (3.9)) up to the largest value of q satisfying:

$$\sum_{p \leq q} \tau_p \leq \Delta t.$$

In the case where $f = 0$, we easily verify that with each time step we have an error of the following type (for the norm $L^\infty(\mathcal{D} \times \mathcal{V})$):

$$\|\bar{u}^{n+1} - \tilde{u}(t^n + \Delta t)\|_\infty = C\Delta t^2 \|\sigma\|_\infty \|\bar{u}^n\|_\infty,$$

where $\tilde{u}$ is the solution of the transport equation whose value in t^n is $\bar{u}^n$. Therefore, if we denote by T the final time and $n(T)$ the number of time steps between 0 and T, we have an estimate of the error in the splitting method:

$$\|\bar{u}^{n(T)} - u(T)\|_\infty \leq C\Delta t \|\sigma\|_\infty \|u(0)\|_\infty.$$

This shows that the splitting method can be legitimately employed if we have a time step Δt, which is small with respect to the inverse of $\|\sigma\|_\infty$.

Remark In practice this method is used:

- For linear equations in cases of weak collisions.
- For nonlinear equations, particularly Boltzmann equations (see Chapter 4), in which case particular care must be taken with the choice of time step (so that the criterion above is satisfied).
- We should note that the error studied here has a different source than errors due to the spatial discretization and to the evaluation of the weights of the particles (see above) or due to a very small number of particles (that is, the quantities are evaluated with a finite sample size instead of a theoretically infinite number).

3.7.3   *Generalization to nonlinear problems*

For certain nonlinear problems, we can use a variant of the classical Monte-Carlo method with spatial discretization. For example, assume that the source depends on the function u and that we must solve the following equation:

$$\frac{\partial u}{\partial t} + v \cdot \frac{\partial u}{\partial x} + ru - Ku = F(\tilde{u}), \qquad (3.21)$$

where F is positive nonlinear function and where we have set:

$$\tilde{u}(x) = \int u(x, v') \, dv'.$$

If $F(U)$ is small with respect to rU, we can evaluate the source explicitly, that is, take, over the time step $[t^n, t^n + \Delta t]$:

$$f = F(\tilde{u^n}).$$

On the other hand, if $F(U)$ is of order greater than rU, we must be careful. If we do not want to have a systematic shift of the source with respect to the evolution of u, we must modify equation (3.21) over the time step $[t^n, t^n + \Delta t]$, for example, in the following way:

$$\frac{\partial u}{\partial t} + v \frac{\partial u}{\partial x} + ru - Ku = \frac{F(\widetilde{u^n})}{\widetilde{u^n}} \int u(v') \, dv'.$$

We then see a second collision operator B appearing:

$$Bu = -ru + \frac{F(\widetilde{u^n})}{\widetilde{u^n}} \int u(v') \, dv',$$

which we can treat in the same way as K. But, as before, this treatment is only possible if we are not in a situation where

$$\bar{v}F(\tilde{u})\tilde{u}^{-1} \ll \Delta\bar{x}.$$

Remark We have seen an example of the solution by linearization of a problem with weak nonlinearity. In this case, there occurs, at each time step, an error due to this linearization and we must analyse this error for every specific problem in order to deduce a criterion on the time step (see, e.g. Alcouffe *et al.*, 1985, for the case of the transport equations of photons).

3.7.4 *Coupling with other numerical methods*

It is possible to use Monte-Carlo methods in one domain of calculation and another method in a neighbouring domain, for example, with a model using the diffusion approximation of the transport equation (in fact, this is common for numerical simulations that use the Boltzmann equation in one domain and the Navier–Stokes equation in another, see Chapter 4). The difficulty is then to construct the coupling condition between the two methods (for the Boltzmann case, see, e.g. Bourgat *et al.*, 1992). On the other hand, for certain models we must consider a system of the following type. We look for two functions $u = u(x, v)$ defined over $\mathcal{D} \times \mathcal{V}$ and $\theta = \theta(t, x)$ defined over $[0, T] \times \mathcal{D}$ and satisfying:

$$v \frac{\partial u}{\partial x} + ru - Ku = \theta,$$

$$\frac{\partial \theta}{\partial t} + F(\theta, \tilde{u}) = 0 \quad \theta(0, x) = \theta_0(x),$$

where θ_0 is given and F is a function which is Lipschitz in its two arguments (in general nonlinear) and $\tilde{u}$ is defined above.

One method to solve this problem numerically is, for example, to make a P_0 discretization in space of the function θ, that is, approximate it in the form:

$$\theta(t, x) = \sum_i \theta_i(t) \xi_i(x),$$

where ξ_i is the indicator function of the element M_i.

We denote by $U_i(x) = \tilde{w}$ where w is the solution of

$$v \frac{\partial w}{\partial x} + rw - Kw = \xi_i \tag{3.22}$$

and make the approximation

$$U_i(x) \simeq \sum_j U_{i,j} \xi_j(x).$$

That is, $U_{i,j} \simeq \int_{M_j} U_i(x) \, dx / |M_j|$. Then, $\tilde{u} \simeq \sum_j U_{i,j} \theta_j$ and the initial problem can then be rewritten in the following semidiscrete form:

$$\frac{\partial \theta_i}{\partial t} + F\left(\theta_i, \sum_j U_{i,j} \theta_j\right) = 0, \quad \theta_i(0) = \theta_{0,i}.$$

We have reduced the problem to a classical one of resolution of the ordinary differential equations in a space of dimension equal to the number of elements in space and to the evaluation of the matrix:

$$\{U_{i,j}\}.$$

Now for this evaluation, we can use a Monte-Carlo method: for each test function ξ_i we solve equation (3.22) using the classical technique for a stationary problem, the quantities $U_{i,j}$ are then obtained by evaluation of the mesh quantities on the elements j. The advantage of this method is that we calculate the matrix $U_{i,j}$ once and for all, then we must treat a simple ordinary differential equation.

This method is called the 'symbolic Monte-Carlo' method, because we can solve the transport equation with real source term, but with a source term taken equal to 1 symbolically in each mesh. This technique has been used, for example, in N'Kaoua and Sentis (1993) in a slightly less general framework where the transport equation has a time derivative.

3.8 Specific techniques

3.8.1 Grouping

In practice, the cross-section σ is often a function which only depends on the velocity through its norm $|v|$, but which can be a very strongly oscillating function of $|v|$. Therefore, the neutron cross-sections must be tabulated as functions of $|v|$ and often we must run a small program to evaluate $\sigma(|v|, x)$.

Now we must know for each particle i and for each element m the value of $\sigma_m(|v_i|)$.

That is why in order to implement the Monte-Carlo method effectively we often resort to a technique called 'grouping', which consists of discretizing the space of velocities $\mathcal{V} = \mathbb{R}^3$.

This technique is only interesting if $\sigma = \sigma(|v|, x)$, and we assume this in what follows. We consider equation (3.7), rewritten in the following form (with $Ku = Lu - \sigma u$):

$$\frac{\partial u}{\partial t} + v \frac{\partial u}{\partial x} + (\sigma + r)u - Lu = 0.$$

The principle is simple:

(a) We discretize the set $\mathbb{R}^+$ of values taken by $|v|$ in G intervals $I_1 I_2 \ldots I_g \ldots I_G$, and we take an average velocity α_g over each I_g. We then replace the continuous velocity v of $\mathbb{R}^3$ by an element $(w, g) \in S \times \{1, 2, \ldots, G\} = S^G$ where S is the unit sphere of $\mathbb{R}^3$. We take

$$w = \frac{v}{|v|}; \quad g \text{ such that } |v| \in I_g.$$

(b) We replace the continuous equation satisfied by $u(t, x, v)$ by a discrete equation satisfied by $u_g(t, x, w)$ (called the multigroup transport equation):

$$\frac{\partial u_g}{\partial t} + \alpha_g w \frac{\partial u_g}{\partial x} + (\sigma_g + r_g)u_g - \sum_{g'} \ell_{g'g}(x) H_{g'g}(u_{g'})(w)$$

with

$$H_{g'g}(u_{g'})(w) = \int_S h_{g'g}(w', w)\sigma_{g'}(x, w')u_{g'}(x, w') \, dw'$$

given that

$$\sum_g \ell_{g'g} = 1, \qquad \int_S h_{g'g}(w', w)\, dw = 1.$$

(c) We adapt the general principles stated in Section 3.4, to the case where $\mathcal{V} = S^G$.

Remark Assume that $r = 0$. We shall show that the system is conservative. Note therefore

$$U = \{u_1, u_2 \ldots u_g\} \in [L^1(S)]^G.$$

We defined the operator A over $[L^1(S)]^G$ by

$$(AU)_g = \sigma_g u_g - \sum_{g'} \ell_{g'g} H_{g'g}(u_{g'}).$$

From the properties above we verify without difficulty that

$$\sum_g \int (Au)_g(w)\, dw = 0.$$

The advantage of this grouping is that at each time step we can store all the values of the cross-section that are used. Indeed, it is enough to store, for each element,

$$G \text{ scalars: } \{\sigma_1 \sigma_2 \ldots \sigma_G\};$$

then we can determine the particles which must jump very quickly (and it will also be easy to vectorize this part of the algorithm).

The difficulty of this technique of grouping is to determine the average σ_g of $\sigma(|v|)$ in the integral I_g, as $\sigma(|v|)$ can oscillate strongly over I_g. We can try the naive formula (arithmetic average):

$$\sigma_g(x) = \frac{1}{|I_g|} \int_{I_g} \sigma(x, z)\, dz.$$

But, let us consider a simple stationary problem:

$$v\frac{\partial u}{\partial x} + \sigma(|v|)u - \int \sigma(|v'|)u(x, v')\, dv' + ru = 0.$$

If $|v|\sigma^{-1}$ is very small compared with the characteristic dimensions then as has been shown above $u(x, v)$ can be approximated by $\Phi(x)\phi(|v|)$ where Φ is the solution of a diffusion equation whose diffusion coefficient $D(x)$ is given by

$$D(x) = \int_0^{+\infty} \frac{\phi(V)}{3\sigma(x,V)} V^3 \, dV.$$

We therefore see that if we only use the arithmetic average σ_g with large intervals I_g (that is, such that σ oscillates often in the interior of I_g) then we shall lose information contained in the data $\sigma(x,\cdot)$. Indeed, the real value of $D(x)$ can be far from $\bar{D}(x)$, the equivalent diffusion coefficient of the multigroup problem which will be

$$\bar{D}(x) = \sum_g I_g \frac{\phi(\alpha_g)\alpha_g^3}{3\sigma_g}.$$

3.8.2 *Fictitious shock technique*

When the average free path $|v|/\sigma(|v|)$ is clearly larger that the size of the elements, it may be interesting to use this technique which allows us to avoid the systematic evaluation of the exact value of $\sigma(|v_i|)$ for each particle i (which can be a very expensive calculation, particularly if we do not use grouping). Let us recall the principle of this technique. We consider problem (3.7) with $f = 0$. Assume that σ only depends on $|v|$ and that there exists a simple function β of $|v|$ satisfying

$$\sigma(v) \leq \beta(v)$$

but such that $|v|\beta(v)^{-1}$ is sufficiently large compared with the size of the elements.

Set

$$\gamma(v) = (\beta(v) - \sigma(v))/\beta(v),$$

then we can replace the initial equation by the following equation:

$$\frac{\partial u}{\partial t} + v\frac{\partial u}{\partial x} + ru + \beta(v)u - Hu = 0 \tag{3.23}$$

with

$$Hu(v) = \beta(v)\gamma(v)u(v) + \int \sigma(v')k(v',v)u(v')\,dv'$$
$$= \int \beta(v')u(v')\mu(dv',v) \tag{3.24}$$

given that $\mu(dv',v)$ is the measure defined by

$$\begin{cases} \mu(dv',v) = (1 - \gamma(v'))k(v',v)dv' + \gamma(v')\delta_v(dv') \\ \text{where } \delta_v(\cdot) \text{ denotes the Dirac mass at } v. \end{cases}$$

We verify immediately that $\mu(A, \cdot)$ is (for every set A) a probability density and the adjoint operator of H will be defined by

$$H_\phi^*(v) = \beta(v) \int [(1 - \gamma(v))k(v, w)\phi(w) + \gamma(v)\delta_v\phi(w)] \, dw.$$

We are therefore in the framework of the previous chapter, since

$$H^*1(v) = \beta(v),$$

$$H^*\phi(v) \geq 0 \quad \forall \phi \geq 0.$$

For the numerical solution, we refer to equation (3.23).

We consider a particle whose velocity is V_i and lifetime is θ_i (before the end of the time step or leaving the element):

- We evaluate a time of 'possible jump' τ as the stopping time chosen according to law with parameter $\beta(V_i)$.
- If $\tau_i \leq \theta_i$, we carry out the 'possible jump', that is
  - with probability $\gamma(v)$ we do not jump;
  - with probability $(1 - \gamma(v))$ we jump according to the measure with density $k(v, w) \, dw$.
- Then we continue the algorithm using the classical scheme.

We easily verify that the probability for the particle with velocity V_i does not have its velocity modified before the time t is $\exp(-\sigma(V_i)t)$.

With this algorithm, it is clear that we only have to look for the value of $\sigma(V)$ for the particles i such that $\tau_i \leq \theta_i$, which is often very small if $V_i\beta(V_i)^{-1}$ is large compared with the size of the elements.

3.9 Reduction of variance and importance functions

Reduction of variance techniques is very important if the values of the unknown function (which is in general a particle density) varies over several orders of magnitude between different parts of the geometry or if the classical method gives highly oscillatory results. The fluctuations are often due to the fact that the masses of the particles are very variable and that there are not enough particles in the 'interesting zones'. One method for reducing the variance of the results is to use the techniques of importance functions. We first make a general remark which is at the root of the technique.

Let $h = h(x, v)$ be a positive function defined over $\mathcal{D} \times \mathcal{V}$. We consider the classical problem:

$$\begin{cases} \dfrac{\partial u}{\partial t} + v\dfrac{\partial u}{\partial x} + ru - Ku = f, \\ u(0, \cdot) = g, \end{cases}$$

with suitable boundary conditions and the notation described in the beginning of the chapter. If we set:

$$\psi(t, x, v) = u(t, x, v)/h(x, v),$$

the preceding problem is equivalent to looking for ψ satisfying:

$$\frac{\partial \psi}{\partial t} + v \frac{\partial \psi}{\partial x} + R\psi - \overline{K}\psi = f/h, \qquad (3.25)$$

where

$$R(x, v) = r(x, v) + \frac{1}{h(x, v)} v \cdot \frac{\partial h}{\partial x} - \overline{\sigma}(x, v) + \sigma(x, v),$$

$$\overline{\sigma}(x, v) = \int \frac{1}{h(x, w)} \sigma(x, v) k(x, v, w) h(x, v) \, dw,$$

$\overline{K}$ a classical conservative operator (of the same form as K) defined by $\overline{\sigma}$ and $\overline{k}$:

$$\overline{k}(x, v', v) = \sigma(x, v') k(x, v', v) h(x, v') / (\overline{\sigma}(x, v') h(x, v)).$$

Remark If h only depends on x then we have:

$$\begin{cases} \overline{K} = K, \overline{\sigma} = \sigma \\ R = r + v \cdot \frac{\partial}{\partial x(\log h)} \end{cases} \qquad (3.26)$$

and if h only depends on v then we have:

$$\begin{cases} \overline{\sigma}(x, v) = \sigma(x, v) h(v) \int k(v, w) \frac{1}{h(w)} \, dw \\ \overline{k}(v', v) = \frac{k(v', v)}{h(v)} \left[\int k(v', w) \frac{1}{h(w)} \, dw \right]^{-1} \\ R(x, v) = r(x, v) + \sigma(x, v)[1 - h(v) \int k(v, w) \frac{1}{h(w)} \, dw] \end{cases} \qquad (3.27)$$

The principle of the technique is simple: we solve equation (3.25) by the classical method of making the approximation:

$$\psi(t, x, v) \, dx \, dv \simeq \sum_i w_i(t) \delta_{X_i(t)} \delta_{V_i(t)}, \qquad (3.28)$$

where (X_i, V_i) are realizations of the Markov process associated with the dual generator of

$$\left(-v \frac{\partial \cdot}{\partial x} + \overline{K} \right)$$

and w_i is the solution for each particle i of

$$\frac{\partial w_i}{\partial t} + R(X_i(t), V_i(t))w_i(t) = 0.$$

Finally, we have:

$$u(t, x, v)\, dx\, dv \simeq \sum_i w_i(t) h(X_i(t), V_i(t)) \delta_{X_i(t)} \delta_{V_i(t)}. \qquad (3.29)$$

There are numerous applications, we cite three.

3.9.1 Angular bias

We can first of all assume that h only depends on v. From formulae (3.27), the replacement of the equation in u by the equation in ψ, corresponds numerically to changing the cross-section (σ becomes $\bar{\sigma}$) and to a change of the law governing the jump in velocity at each collision (k becomes $\bar{k}$).

From the viewpoint of the Monte-Carlo method, this reduces to a biased choice of the velocity at each collision. We can use this technique, for example, in the following case: the physical source of particles is in a given zone A and we are interested in the solution on another zone B (far from the first), it is then interesting to bias the choice of the velocity at each collision so that it is principally oriented from A to B.

Suppose therefore that

- The cross-section σ only depends on x and $|v|$.
- The kernel only depends on x and $|v|$, $|v'|$.
- We want to give greater importance to the velocities whose direction belong to the segment of the sphere Σ such that the ratio between the surface of Σ and that of S^2 is δ.

Therefore, we define the importance function in the following way (we write $\vec{v} = |\vec{v}|\Omega$ where $\Omega \in S^2$):

$$h(v) = h(\Omega) = \begin{cases} 1 & \text{if } \Omega = v/|v| \in \Sigma^C, \\ M^{-1} & \text{if } \Omega \in \Sigma, \end{cases}$$

we then have:

$$\bar{\sigma}(x, v) = \sigma(x, |v|) h(\Omega)(1 - \delta + \delta M),$$
$$\bar{k}(x, v', v) = \frac{k(x, |v'|, |v|)}{h(\Omega)}(1 - \delta + \delta M)^{-1},$$
$$R(x, v', v) = r(x, v) + \sigma(x, |v|)(1 - h(\Omega)(1 - \delta + \delta M)).$$

We take $M > 1$, for the choice of the angle Ω of the velocity after each collision, we continue in the following way:

Ω uniformly distributed over Σ with probability $\quad \dfrac{\delta M}{1 - \delta + \delta M}$

Ω uniformly distributed over Σ^c with probability $\quad \dfrac{1 - \delta}{1 - \delta + \delta M}$.

3.9.2   Source bias

We can choose a function h only depending on x and differentiable in x, we then have:

$$R(x, v) = r(x, v) + v \cdot \frac{\partial}{\partial x}(\log h).$$

To evaluate the function ψ, one then operates the same trajectory calculation on the particles as that which would have been done for the function u; on the other hand, the variation of the particle weights is different. Therefore, if we return to the initial function u using formula (3.29), we change nothing, in practice, from the classical method, since we do not change the trajectory calculation.

The interest in this weight bias lies uniquely in the fact that it is used in conjunction with a technique of threshold weights (in which we destroy the particles that have weights less than a certain threshold so that we do not have to follow too many particles).

3.9.3   Geometry splitting

We can choose the function h constant in each zone. Therefore, assume that we perform a partition of $\mathcal{D}$ into two subsets $\mathcal{D}_1$ and $\mathcal{D}_2$ where the common boundary is denoted Γ. Let M be a constant greater than 1, set:

$$h(x) = \begin{cases} 1 & \text{if } x \in \mathcal{D}_1, \\ M^{-1} & \text{if } x \in \mathcal{D}_2. \end{cases}$$

We then have:

$$R(x, v) = r(x, v) - \delta_\Gamma n_\Gamma \cdot v \log M,$$

where n_Γ is the normal to Γ directed from $\mathcal{D}_1$ to $\mathcal{D}_2$. δ_Γ is the uniform surface measure on Γ of mass 1. When we solve the equation:

$$\frac{\partial w_p}{\partial t} + R w_p = 0,$$

the introduction of the Dirac mass comes down to multiplying the weights w_p by M when the particle p crosses Γ and $n_\Gamma \cdot V_p$ is positive.

In practice, to evaluate the function ψ we carry out the following (called Russian Roulette and Splitting):

- When a particle crosses Γ from $\mathcal{D}_1$ to $\mathcal{D}_2$, we replace it by M particles having the same weight, the same velocity (if M is not an integer, we choose at random between $[M]$ and $[M] + 1$ with suitable probabilities).
- When a particle crosses Γ in the other direction we destroy it with a probability M^{-1}.

3.10 An example of angular bias

We consider here a simple transport equation in two dimensions; the spatial domain is the following:

$$\mathcal{D} = \{(x_1, x_2) | 0 < x_1 < 15;\ 0 < x_2 < 17\}$$

and the velocity domain is

$$\mathcal{V} = \{v \in \mathbf{R}, |v| \le 1\},$$

equipped with the uniform measure. We look for $u(x, v)$ the solution of the stationary equation:

$$\begin{cases} v \cdot \dfrac{\partial u}{\partial x} + ru + \sigma(u - \displaystyle\int u(v')\,dv') = 0, & x \in \mathcal{D},\ v \in \mathcal{V}, \\ u(x, v) = g(x)G(v), & x \in \partial\Omega,\ v \in Z_x, \end{cases}$$

where g is the indicator of the segment $(x_1 = 0,\ x_2 \in [2, 7])$ and G a positive function whose support is concentrated close to $v = (v_1, v_2) = (0, 1)$ and whose integral is 1.

We take:

$$r = 0.1, \qquad \sigma = 0.9.$$

We want to evaluate the outward flux $F(x_2)$ on the boundary ($x_1 = 15$) as a function of x_2. Two calculations have been made, one by the classical method, the other by using biasing by the following importance function:

$$h(v) = \frac{1}{1 - kv_1},$$

where k is a positive constant, which, after experimentation, has been taken equal to 0.52 (this choice of importance function will favour the displacement of particles in the direction x_1, and must therefore improve the statistics of the answer sought). We show, in Figs 3.1 and 3.2, the results obtained for the function $F(x_2)$ with the two methods, by generating the same number of particles: 200 000. The improvement in the statistics is spectacular with the biased method.

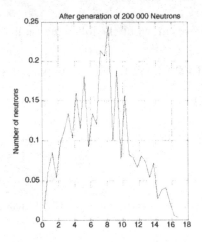
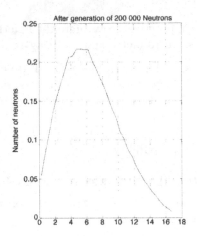

Fig. 3.1. Simulation without bias.   **Fig. 3.2.** Simulation with bias.

3.11   Remarks about programming

Effective programming relies on the management of the tables containing the characteristics of the particles. In general, for problems of evolution in time, we generate particles at all the timesteps. So that the number of particles does not become too large, we destroy the particles when their weight becomes lower than a certain threshold depending on the domain where they are and on their initial weight.

The adaptation of Monte-Carlo methods for transport, to modern computer architecture, is a vast subject. We shall only give some ideas about vectorization and parallelization.

3.11.1   *Vectorization*

The problem of vectorization of Monte-Carlo methods is classical, but nonetheless requires significant work to program.

We must treat the particles by packet, but in each packet the particles have different histories.

Schematically, using the notation introduced in the algorithm described in this chapter, for each particle of a packet we will have:

1. evaluate the three associated times: t_q, τ_q, θ_q;
2. calculate $s = \min(t_q, \tau_q, \theta_q)$;
3. make three lists of particles (at least, since we can make sublists for some particular boundary conditions) corresponding to three events:

$$s = t_q,$$
$$s = \tau_q,$$
$$s = \theta_q,$$

4. After modification of the parameters of the particles in the first two cases and suppression of the particles in the third case, we return to point 1 (until we have exhausted all the particles in a packet).

We see that we have to make intense use of lists of particles satisfying certain criteria and of indirect addressing.

3.11.2 *Parallelization*

If we consider the parallelization of these methods on MIMD machines (Multiple Instructions, Multiple Data) with shared memory, two important strategies take shape:

- We shall distribute the processors to packets of particles. This is possible if, on the one hand, we can store all the useful geometrical and physical data on each processor and if, on the other hand (in the case of time dependent problems), it is not necessary to estimate the grid quantities ϕ and J at each time step (since the estimation of these quantities is global and would need many transfers of information).

- We can distribute the processors to the subdomains of the grid; in this case, we must control the transfer of particles from one subdomain to another. We should distinguish two kinds of problems:

  - Problems where the time discretization does not intervene. That is, the case of the stationary problems and of evolution problems with coefficients independent of time.

    There is no conceptual difficulty. We assign, to each processor, one and only one subdomain and in each of the subdomains D, we treat the particles as in D up until they all leave the subdomain (or their lifetime is reached in which case they are destroyed), we then transfer the information between the processors and repeat the process.

  - Problems where time discretization is crucial. This is the case of evolution problems with cross-sectional coefficients that change in the course of time (or, *a fortiori*, coefficients that depend on the solution at the preceding time step).

    The general principle is to work at time step in the following way: For each subdomain D, we work with the particles contained in D at the beginning of the timestep, then we transfer information between processors, then for each D we work with the particles that have entered D during time step. Theoretically, we should then carry out transfers of information for particles that could have belonged to three subdomains during the timestep. These situations are very infrequent but are very costly in terms of calculation, the difficulty is to find an inexpensive solution for these situations.

For these two types of problem, the balancing of tasks between the various processors is a significant and delicate subject.

3.12 Bibliographic comments and conclusions

We have not detailed here numerical methods related to the adjoint Monte-Carlo method; they are of the same type as those we have presented (see Hockney and Eastwood, 1981; Alcouffe *et al.*, 1985).

There is much work on the reduction of variance, some of these techniques are very heuristic, others are only suitable for a particular problem; in any event, the techniques of reduction of variance must 'follow' the physics of the problem to be effective. For techniques that can diminish the statistical fluctuation and ensure an effective implementation, see, for example, Spanier and Gelbard, (1969), Alcouffe *et al.*, (1985), and Booth (1985).

The methods that we have just described have the advantage of being very flexible; for example, one can start from a rather simple model with only one type of collision operator then enrich the model by taking account of several different collision operators or to stop the numerical treatment of a collision operator in a zone if we know that it is not significant.

In addition, one can deal with very large problems without too many difficulties by adjusting the number of particles (and therefore the cost of calculation) with the desired precision. We remark finally that the precision of the method, and therefore its interest, is large if we are only looking for a small number of scalar results (e.g. the value of the solution in a small part of the domain or on the boundary) and we are then able to use the methods of reduction of variance that are able to support precisely and easily the evaluation of results in a given domain.

4

The Monte-Carlo method for the Boltzmann equation

The Boltzmann equation describes the evolution of a population of particles that have binary collisions and travel in a straight line between each collision. This type of equation can be used in a great number of physical models where we want to describe collision phenomena carefully. A significant example of the application is the modelling of the flows around objects in the upper atmosphere (above 70 km) where the distribution of the particles is not necessarily a Maxwellian function. This equation can be also used to model the behaviour of low-density gas in laboratory experiments.

For a long time, Monte-Carlo methods were the only ones used to solve the Boltzmann equation but they were introduced historically as being a direct simulation (with a restricted number of particles) of the underlying statistical mechanics process, namely the dynamics of a rarefied gas comprising a very large number of particles (a number such that the particle distribution function has deterministic behaviour), from where we get the name of the method: 'Direct Simulation Monte-Carlo' by Bird (1963).

The process will be different here from that of the preceding chapters on the linear transport equation; we will not try to show, on the complete equation, results of the type: 'the solution of the Boltzmann equation is the mathematical expectation of a random process' (indeed, such a rigorous and complete result does not yet exist, see however the bibliographical comments at the end of the chapter concerning work in this direction). Nevertheless, we will establish a rigorous relation between the spatially homogeneous Boltzmann equation and the limit as $N \to \infty$ of 'the principal equation of a system with N particles', which is a Fokker–Planck equation for a Markov process in the space R^{3N} (which allows us to give an interpretation in terms of a Monte-Carlo method). The principle of all the numerical methods is initially to perform a splitting between the advection part and the collision part. The treatment of the collision phase can be carried out by two types of methods: linear or symmetric. For the first, we linearize the collision operator at each timestep and we act by analogy with what we did for the transport equation. For the second, we keep the symmetry of the collisions and recall the probabilistic interpretation of the spatially homogeneous equation mentioned above.

It is important to note that we cannot make collisions 'in flight' as for the transport equation, indeed the cross-section 'in flight' is unknown, as it depends on a distribution at the point of space considered. This is why the splitting between the advection part and the collision part is essential, contrary to what was shown in preceding chapters.

4.1 General points about the Boltzmann equations

We restrict ourselves here to the most simple Boltzmann equations:

- for only one type of particle;
- for particles not having internal energy.

However, most of the expositions that we shall make can be adapted to more general situations. Conforming to tradition, we denote by $f(t, x, v)$ the population of particles at the time t having velocity v (belonging to $\mathbf{R}^3$) and position x (belonging to an open set $\mathcal{D}$ of $\mathbf{R}^3$). The Boltzmann equation is then written as

$$\frac{\partial f}{\partial t} + v \cdot \frac{\partial f}{\partial x} = Q(f, f), \tag{4.1}$$

where Q is the collision operator:

$$Q(f, f)(v) = \int_{\mathbf{R}^3} \int_{S^2} q\left(v - v_1, \sigma \cdot (v - v_1)\, |v - v_1|^{-1}\right) \\ \times \left[f(v')f(v_1') - f(v)f(v_1)\right] d\sigma\, dv_1$$

given that $\sigma \in S^2$ (the unit sphere of $\mathbf{R}^3$) and that we have set:

$$v' = \tfrac{1}{2}(v + v_1) + \tfrac{1}{2}\sigma\, |v - v_1|, \qquad v_1' = \tfrac{1}{2}(v + v_1) - \tfrac{1}{2}\sigma\, |v - v_1|. \tag{4.2}$$

The measure $d\sigma$ is the unit measure over S^2 and q is a positive function defined over $\mathbf{R}^3 \times [-1, 1]$ such that $q(w, \mu)$ only depends on $|w|$ (the relative velocity between the particles) and μ (the cosine of the angle between the relative velocities pre- and post-collision). The quantity $q_0(w, \mu) = q(w, \mu)/|w|$ is called the microscopic cross-section. The most simple, and most common, cases are

$$q_0(w, \mu) = C, \quad q(w, \mu) = C\, |w|,$$

$$q_0(w, \mu) = C\, |w|^{\alpha - 1}, \quad q(w, \mu) = C\, |w|^\alpha, \qquad \alpha > 0,$$

which are called, respectively, 'Hard Spheres' and 'Variable Hard Spheres'.

We can define an average free time for collisions corresponding to a particle of velocity v by

$$1 \Big/ \int \left(\int_{-1}^{1} q(v - v_1, \mu)\, d\mu/2 \right) f(v_1)\, dv_1.$$

In the case where $\mathcal{D}$ is not the whole space, we must give conditions on the boundary of $\mathcal{D}$ in the same way as has been done for the transport equation in Chapter 3.

We see that we have conservation of momentum and of kinetic energy for the shock conditions defined by (4.2), that is,

$$v' + v'_1 = v + v_1,$$
$$|v'|^2 + |v'_1|^2 = |v|^2 + |v_1|^2 .$$

We give here some elementary properties of the operator Q, which allow us to write the fundamental relations of global conservation of momentum and kinetic energy (which must also be satisfied by the numerical scheme).

The quadratic operator $Q(f, f)$ induced by the following bilinear form (which we use to linearize the problem):

$$Q(f, g) = -L(g)f + S(f, g),$$

where the linear operator L and the quadratic operator S are given by

$$L(g) = \int_{\mathbf{R}^3} \int_{S^2} q(v - v_1, \sigma.(v - v_1)|v - v_1|^{-1})g(v_1)\, dv_1\, d\sigma,$$

$$S(f, g)(v) = \int_{\mathbf{R}^3} \int_{S^2} q(v - v_1, \sigma.(v - v_1)|v - v_1|^{-1})f(v')g(v'_1)\, dv_1\, d\sigma,$$

given that v' and v'_1 are given by (4.2).

Without entering into detail about the function spaces in which we find f and g, we state simply that if $q(w, \cdot)$ is an increasing polynomial in $|w|$, then f and g must 'decrease rapidly to infinity' in order that $Q(f, g)$ be, for example, in a weighted $L^1(\mathbf{R}^3)$ space.

For fixed g, the operator $Q(f, g)$ can be interpreted as modelling the effect of collisions of the population represented by g with the population represented by f.

Proposition 4.1.1 *For every function f and g defined over $\mathbf{R}^3$ and 'decreasing rapidly to infinity', we have:*

(i) $\displaystyle\int_{\mathbf{R}^3} Q(f, g)(v)\, dv = 0;$

(ii) $\displaystyle\int_{\mathbf{R}^3} vQ(f, f)(v)\, dv = 0;$

(iii) $\displaystyle\int_{\mathbf{R}^3} v^2 Q(f, f)(v)\, dv = 0.$

Proof By making the change of variables

$$v, v_1, \sigma \rightarrow v', v'_1, \sigma',$$

where $\sigma' = (v - v_1)|v - v_1|^{-1}$, and remarking that:

$$q(v' - v'_1, \sigma' \cdot (v' - v'_1)|v' - v'_1|^{-1}) = q(v - v_1, \sigma \cdot (v - v_1)|v - v_1|^{-1}),$$

we verify that for every function ϕ defined over $\mathbf{R}^3$, we have:

$$\int \phi(v) S(f,g)(v)\, dv$$

$$= \int \int \int_{S^2} \phi(v) q(v - v_1, \sigma \cdot (v - v_1) \, |v - v_1|^{-1}) f(v') g(v_1') \, dv_1 \, dv \, d\sigma$$

$$= \int \int \int_{S^2} \phi(v) q(v' - v_1', \sigma' \cdot (v' - v_1') \, |v' - v_1'|^{-1}) f(v') g(v_1') \, dv_1' \, dv' \, d\sigma'.$$

By making a change of notation (v', v_1', σ' becomes v, v_1, σ and vice versa), we deduce from the last relation that

$$\int \phi(v) S(f,g)(v)\, dv$$

$$= \int \int \int \phi(v') q(v - v_1, \sigma.(v - v_1) \, |v - v_1|^{-1}) f(v) g(v_1) \, d\sigma \, dv \, dv_1.$$

From which we have result (i) by setting $\phi = 1$.

For points (ii) and (iii), we write an analogous relation with v' changed to v_1':

$$\int \phi(v) Q(f,f)(v)\, dv$$

$$= \frac{1}{2} \int \int \int [\phi(v') + \phi(v_1') - \phi(v) - \phi(v_1)]$$

$$\times q(v - v_1, \sigma \cdot (v - v_1) \, |v - v_1|^{-1}) f(v) f(v_1) \, d\sigma \, dv \, dv_1 \qquad (4.3)$$

and we choose successively:

$$\phi(v) = v \quad \text{and} \quad \phi(v) = v^2.$$

The result is the simple consequence of the conservation relations:

$$\phi(v) + \phi(v_1) = \phi(v') + \phi(v_1').$$

$\square$

Therefore, if we do not take account of the dependence on x, the solution $f = f(t,v)$ of the spatially homogeneous equation:

$$\frac{\partial f}{\partial t} = Q(f,f)$$

is such that we have conservation of the total number of particles

$$\int f(v)\, dv,$$

with total momentum

$$\int v f(v) \, dv,$$

and with total kinetic energy

$$\int \frac{v^2}{2} f(v) \, dv.$$

We can also show a decrease in the mathematical entropy

$$\int f(v) \, \log \, f(v) \, dv,$$

by using the property

$$\int Q(f, f)(v) \, \log \, f(v) \, dv \le 0.$$

For a general approach to the Boltzmann equations, see, for example, Cercignani (1988). For completeness and though it is not relevant to our problem, we give two results below on the existence of solutions of the Boltzmann equation. On the one hand, in the spatially homogeneous case and under the assumption (which we will make from now on):

$$q(v - v_1) \le C \, |v - v_1|^{\alpha} \, .$$

Denoting by $L_2^1(\mathbf{R}^3) = \{f \text{ s.t. } \int f(v)(1 + |v|^2) \, dv < \infty\}$, we have the following result, see Arkeryd (1981) and Mischler and Wennberg (1996).

Proposition 4.1.2 *If we make the hypothesis that the initial condition f^0 is positive and satisfies:*

$$f^0 \in L_2^1(\mathbf{R}^3),$$

$$\int f^0(v) \, \log \, f^0(v) \, dv < \infty.$$

Then, there exists a continuous unique solution from $\mathbf{R}^+$ with values in $L_2^1(\mathbf{R}^3)$ of the spatially homogeneous equation $\partial f / \partial t = Q(f, f)$.

On the other hand, the proof of an existence result for a global solution is recent and technical see Di Perna and Lions (1989). This result can be stated in the following form in the case where the domain $\mathcal{D}$ is $\mathbf{R}^3$, by setting $L_M^1(\mathbf{R}^3 \times \mathbf{R}^3) = \{f \text{ s.t. } \int \int f(v) e^{-|v|^2/2} \, dx \, dv < \infty\}$:

Proposition 4.1.3 *If we make the hypothesis that the initial condition f^0*

is positive and satisfies:

$$f^0 \in L^1_M(\mathbf{R}^3 \times \mathbf{R}^3),$$

$$\int \int f^0(v) \log f^0(v) \, dv \, dx < \infty.$$

Then, there exists a solution of equation (4.1) in the sense that we have the following relation (in the space $L^1_{loc}[0, +\infty; L^1_M(\mathbf{R}^3 \times \mathbf{R}^3)]$):

$$\left(\frac{\partial}{\partial t} + v \cdot \frac{\partial}{\partial x} \right) (\log(1+f)) = \frac{Q(f,f)}{1+f}$$

and further f is weakly continuous from $\mathbf{R}^+$ to $L^1_M(\mathbf{R}^3 \times \mathbf{R}^3)$.

4.2   Link with the principal equation

The purpose of this section is to justify a family of Monte-Carlo methods (symmetric method, Bird's method) and more precisely the collision phase. We put ourselves in the framework of spatial homogeneity; it is a question of showing how this method makes it possible to approximate, on a fixed time interval, the solution of the equation:

$$\frac{\partial f}{\partial t} = Q(f,f), \tag{4.4}$$

$$f(0,v) = f^0(v),$$

with $f^0 \geq 0$. Denoting:

$$\rho = \int f^0(v) \, dv, \qquad \hat{f}(t,v) = f(t,v)/\rho. \tag{4.5}$$

From Proposition 4.1.1, $\hat{f}(t)$ is a probability density, for all t. We shall approximate $\hat{f}(t)$ by a sequence of empirical probabilities (that is, a linear combination of Dirac masses), which will be constructed from a jump Markov process with values in $\mathbf{R}^{3N}$ for large N.

First, we make precise the following notation:

- E denotes the space of velocities $\mathbf{R}^3$;
- $C_b(E)$ denotes the space of bounded continuous functions over E and $\langle ., . \rangle$ the duality paring between the measures over E and $C_b(E)$;
- $E^{(N)}$ denotes the Nth symmetric power of E (that is the two elements are equal if we can pass from one to the other by a permutation of their components);
- $\mathbf{v} = (v_1, v_2, \ldots, v_i, \ldots, v_N)$ an element of $E^{(N)}$.

We now introduce the notion of chaotic random variables. Let $\mathbf{V}^N = (V^N_1, V^N_2, \ldots, V^N_i, \ldots, V^N_N)$ be a sequence of random variables with values in $E^{(N)}$ and let μ_0 be a probability over E. For every N, the random variable $\mathbf{V}^N$ with values in $E^{(N)}$ has for a law, a measure μ^N over $E^{(N)}$ (that is, a symmetric measure over E^N). To the random variable $\mathbf{V}^N$, we associate the random measure:

$$\overline{\mathbf{V}}^N = \frac{1}{N} \sum_{i=1}^{N} \delta_{V_i^N}, \quad \delta \quad \text{denotes the Dirac measure over} \quad E.$$

The random measure $\overline{\mathbf{V}}^N$ is called the empirical probability associated with the variable $\mathbf{V}^N$.

Definition 4.2.1 We say that the random variables $\mathbf{V}^N$ are chaotic for the probability measure μ_0 if the random measure $\overline{\mathbf{V}}^N$ converges strictly in mean to μ_0, as $N \to \infty$, that is,

$$\mathbf{E}\left(\left| \langle \overline{\mathbf{V}}^N, \phi \rangle - \langle \mu_0, \phi \rangle \right| \right) \to 0 \quad \text{for all} \quad \phi \in C_b(E).$$

Remark As ϕ is bounded, the convergence in mean is equivalent to convergence in probability, that is, when $N \to \infty$:

$$\mathbf{P}\left(\left| \langle \overline{\mathbf{V}}^N, \phi \rangle - \langle \mu_0, \phi \rangle \right| \geq \quad \epsilon \right) \to 0 \quad \text{for all } \phi \in C_b(E).$$

We can show the following result (see, e.g. Wagner, 1992).

Proposition 4.2.2 *The random variables $\mathbf{V}^N$ are chaotic for the probability measure μ_0 if and only if for every integer j and for all $\phi_1, \phi_2, \ldots \phi_j$ in $C_b(E)$, we have:*

$$\int \int \ldots \int \phi_1(y_1)\phi_2(y_2) \ldots \phi_j(y_j) \mu_N \big|_{E^j} (dy_1\, dy_2 \ldots dy_j) \to \prod_{k=1}^{j} \langle \mu_0, \phi_k \rangle$$

as $N \to \infty$.

Remark Here is an important example of chaotic random variables. Assume that we have a sequence of equidistributed independent random variables $\{W_1, W_2, \ldots, W_i, \ldots\}$ with law μ_0, then the variables $\mathbf{V}^N = \{W_1, W_2, \ldots, W_N\}$ are chaotic for the measure μ_0; indeed, from the law of large numbers, we have as N tends to ∞:

$$\langle \overline{\mathbf{V}}^N, \phi \rangle = \frac{1}{N} \sum_{i=1}^{N} \phi(W_i) \to \langle \mu_0, \phi \rangle \quad \text{almost surely}$$

(the limit being the average of $\phi(W_1)$).

Remark If the random variables $\mathbf{V}^N$ are chaotic for μ_0, then the marginal law of V_1^N converges strictly to μ_0 (this follows from Proposition 4.2.2 and from the fact that the law of $\mathbf{V}^N$ is symmetric).

4.2.1 *Principal equation and the Bird collision process*

We now consider the following linear equation satisfied by the function $h_N = h_N(t, \mathbf{v})$ defined over $E^{(N)}$:

$$\frac{\partial}{\partial t} h_N(\mathbf{v}) = \frac{\rho}{N} \sum_{1 \leq i \neq j \leq N} \int_{S^2} q(g_{i,j}(\mathbf{v}))(h_N((\mathbf{v})'_{i,j}) - h_N(\mathbf{v}))\, d\sigma,$$

$$h_N(0, \mathbf{v}) = \prod_{i=1}^{N} \hat{f}(0, v_i),$$

given that for all $\mathbf{v}$, we use the following notation (the parameter σ being suppressed):

$$(\mathbf{v})'_{i,j} = (v_1, \ldots, v'_i, \ldots, v'_j, \ldots v_N)$$

with v'_i, v'_j given by formulae (4.2) where the parameter σ appears and

$$g_{i,j}(\mathbf{v}) = v_i - v_j.$$

It is easy to verify that for all t, $h_N(t, \cdot)$ is a function defined over $E^{(N)}$ (that is, symmetric over E^N). For simplicity, we assume that $q(w, \mu)$ does not depend on μ. This equation is called the principal equation for a collisional system of N particles. This is a Fokker–Planck equation for the jump Markov process:

$$\mathbf{V}_t^N = (V_{1t}^N, V_{2t}^N, \ldots, V_{it}^N, \ldots, V_{Nt}^N)$$

with values in $E^{(N)}$ defined below:

- The initial condition is such that $(V_{10}^N, V_{20}^N, \ldots V_{i0}^N, \ldots V_{N0}^N)$ are distributed independently according to the density $\hat{f}(0)$ over E.
- The time of the first jump is an exponential random variable with parameter:

$$\frac{\rho}{N} \sum_{i,j=1}^{N} q(g_{i,j}(\mathbf{V}^N)),$$

then with probability

$$\frac{q(g_{i,j}(\mathbf{V}^N))}{\sum_{k,l=1}^{N} q(g_{k,l}(\mathbf{V}^N))}$$

the pair (i, j) is selected and the process jumps from position $\mathbf{v}$ to position $(\mathbf{v})'_{i,j}$ defined above for σ equidistributed over the unit sphere.

The process $\mathbf{V}_t^N$ is called a Bird collision process. We can immediately verify that the law of this process is symmetric with respect to all its arguments.

We remark that the Kolmogorov equation associated with this process is identical to its Fokker–Planck equation, since the infinitesimal generator of the process is symmetric. The aim of this section is to show that we can approximate the solution of equation (4.4) with the help of this process. For this, we must first state an important result on the propagation of chaos.

4.2.2 Propagation of chaos

From the remark given above, the initial value of the Bird collision processes $\mathbf{V}_t^N|_{t=0}$ are chaotic for the measure $\hat{f}(0, v) \, dv$. We have:

Theorem 4.2.3 (Propagation of Chaos) *For every positive t, the $\mathbf{V}_t^N$ are chaotic for the measure $\hat{f}(t, v) \, dv$ (where $\hat{f}$ is given by (4.5)).*

This theorem, whose statement is due to Kac (1956) is tricky to prove rigorously, the first proof seems to be due to Grunbaum (1971). We shall only give here the principle of this proof, which relies on the following lemma:

Lemma 4.2.4 *Let Ψ be a function on $C_b(E^{(2)})$ and let μ and μ^1 be probability measures on E we have:*

$$\left| \int \int \Psi(v, v_*) \mu^1(dv) \mu^1(dv_*) - \int \int \Psi(v, v_*) \mu(dv) \mu(dv_*) \right|$$
$$\leq 2 \sup_v \left| \langle \mu^1 - \mu, \Psi(v, \cdot) \rangle \right|.$$

Proof This lemma follows easily from the fact that ψ is symmetric and from the identity:

$$\int \int \Psi(v, v_*) \mu^1(dv) \mu^1(dv_*) - \int \int \Psi(v, v_*) \mu(dv) \mu(dv_*)$$
$$= \int \langle \mu^1 - \mu, \Psi(., v_*) \rangle \mu^1(dv_*) + \int \langle \mu^1 - \mu, \Psi(v, .) \rangle \mu(dv).$$

$\square$

Proof We only give an outline of the proof. It is inspired by Wagner, (1992). Let ϕ be a function on $C_b(E)$. We can associate with it a function G defined over $E^{(N)}$ by

$$G(\mathbf{V}^N) = \langle \overline{\mathbf{V}}^N, \phi \rangle = \frac{1}{N} \sum_{k=1}^{N} \phi(V_k^N).$$

Using the Kolmogorov equation associated with the jump process $\mathbf{V}_t^N$, we then have (reasoning in the same way that we have in Chapter 2 for the jump processes):

$$\frac{d}{dt}\mathbf{E}(G(\mathbf{V}_t^N)) = \mathbf{E}\left[\frac{\rho}{N}\sum_{1\le i\ne j\le N}\int_{S^2} d\sigma\, q(g_{i,j}(\mathbf{V}_t^N))(G((\mathbf{V}_t^N)'_{i,j})-G(\mathbf{V}_t^N))\right].$$

Since we know that

$$G((\mathbf{V}^N)'_{i,j}) - G(\mathbf{V}^N) = \frac{1}{N}[\phi, ((V_i^N)') + \phi((V_j^N)') - \phi(V_i^N) - \phi(V_j^N)],$$

by using the natural notation for $(._i)'$ and $(._j)'$, we deduce that, for all positive T:

$$\mathbf{E}(G(\mathbf{V}_T^N)) = \mathbf{E}(G(\mathbf{V}_0^N)) + \int_0^T \mathbf{E}\left[\frac{\rho}{N^2}\sum_{i\ne j}\int q(g_{i,j}(\mathbf{V}_t^N))(\phi((V_{i,t}^N)')\right.$$

$$\left. + \phi((V_{j,t}^N)') - \phi(V_{i,t}^N) - \phi(V_{j,t}^N))\,d\sigma\right]dt.$$

Now set:
$$\psi_\sigma(v_1, v_2) = [\phi(v_1') + \phi(v_2') - \phi(v_1) - \phi(v_2)].$$

Now we have:

$$\int\int q(v_1, v_2)\psi_\sigma(v_1, v_2)\overline{\mathbf{V}}_t^N(dv_1)\overline{\mathbf{V}}_t^N(dv_2)$$

$$= \frac{1}{N^2}\sum_{i\ne j} q(g_{i,j}(\mathbf{V}_t^N))(\phi((V_{i,t}^N)') + \phi((V_{j,t}^N)') - \phi(V_{i,t}^N) - \phi(V_{j,t}^N)).$$

We can therefore write:

$$\mathbf{E}(\langle\overline{\mathbf{V}}_T^N,\phi\rangle) = \mathbf{E}(\langle\overline{\mathbf{V}}_0^N,\phi\rangle)$$

$$+ \rho\mathbf{E}\left[\int_0^T\int\int\int q(v_1 - v_2)\psi_\sigma(v_1, v_2)\overline{\mathbf{V}}_t^N(dv_1)\overline{\mathbf{V}}_t^N(dv_2)\,d\sigma\,dt\right].$$

On the other hand, from the weak form of the collision operator (see (4.3)), we know that $\hat{f}$ satisfies:

$$\int \hat{f}(T, v)\phi(v)\,dv = \int \hat{f}(0, v)\phi(v)\,dv$$

$$+ \rho\int_0^T\int\int\int q(v_1, v_2)\psi_\sigma(v_1, v_2)\hat{f}(t, v_1)\hat{f}(t, v_2)\,dv_1\,dv_2\,d\sigma\,dt.$$

By abuse of notation, we denote by $\overline{\mathbf{V}}_t^N - \hat{f}(t)$ the measure $\overline{\mathbf{V}}_t^N - \hat{f}(t, v)\, dv$. Denote also, for every measure μ on E:

$$\|\mu\| = \sup_{\phi \in C_b(E)} \left(|\langle \mu, \phi \rangle| / \|\phi\| \right).$$

From above, we see that

$$\mathbf{E}\left(\langle \overline{\mathbf{V}}_T^N - \hat{f}(T), \phi \rangle \right)$$

$$= \mathbf{E}\left(\langle \overline{\mathbf{V}}_0^N - \hat{f}(0), \phi \rangle \right) + \rho \mathbf{E}\left[\int_0^T \int \int \int q(v_1 - v_2) \psi_\sigma(v_1, v_2) \right.$$

$$\left. \times \left\{ \overline{\mathbf{V}}_t^N(dv_1)\overline{\mathbf{V}}_t^N(dv_2) - \hat{f}(t, v_1)\hat{f}(t, v_2)\, dv_1\, dv_2 \right\} d\sigma\, dt \right].$$

Suppose now that q is bounded by a constant q_1. By using Lemma 4.2.4, we have:

$$\mathbf{E}\left(\left| \langle \overline{\mathbf{V}}_T^N - \hat{f}(T), \phi \rangle \right| \right) \leq \mathbf{E}\left(\left| \langle \overline{\mathbf{V}}_0^N - \hat{f}(0), \phi \rangle \right| \right)$$

$$+ 2q_1 \rho \mathbf{E}\left[\int_0^T \int \sup_v \left| \langle \overline{\mathbf{V}}_t^N - \hat{f}(t), \psi_\sigma(v, \cdot) \rangle \right| d\sigma\, dt. \right.$$

Since we know that $|\psi_\sigma(v, .)| \leq 4\,|\phi|$, we can deduce:

$$\mathbf{E}\left(\left\| \overline{\mathbf{V}}_T^N - \hat{f}(T) \right\| \right) \leq \mathbf{E}\left(\left\| \overline{\mathbf{V}}_0^N - \hat{f}(0) \right\| \right) + 8q_1 \rho \mathbf{E}\left[\int_0^T \left\| \overline{\mathbf{V}}_t^N - \hat{f}(t) \right\| dt. \right.$$

From the Gronwall lemma, we deduce easily that there exists a constant C such that for all t:

$$\mathbf{E}\left(\left\| \overline{\mathbf{V}}_t^N - \hat{f}(t) \right\| \right) \leq \mathbf{E}\left(\left\| \overline{\mathbf{V}}_0^N - \hat{f}(0) \right\| \right) e^{Ct}.$$

Since $\mathbf{V}_0^N$ is chaotic for $\hat{f}(0, v)\, dv$, we then deduce that for all $\phi \in C_b(E)$, we have for $N \to \infty$:

$$\mathbf{E}\left(\left| \langle \overline{\mathbf{V}}_t^N - \hat{f}(t), \phi \rangle \right| \right) \to 0.$$

Moreover, in the general case where q is not bounded, we must use more careful estimates, in particular, we must introduce a ball of fixed radius R, to estimate the probability that the process $\mathbf{V}_t^N$ leaves the ball; and the integral

$$\mathbf{E}\left(\int \int q(v_1 - v_2)\psi_\sigma(v_1, v_2) \left\{ \overline{\mathbf{V}}_t^N(dv_1)\overline{\mathbf{V}}_t^N(dv_2) - \hat{f}(t, v_1)\hat{f}(t, v_2)\, dv_1\, dv_2 \right\} \right)$$

must then be split into two parts, one for which $\mathbf{V}_t^N$ is in the ball, the other for which $\mathbf{V}_t^N$ is outside. $\qquad \Box$

Remark Denoting by

$$h_N^1(t, v) = \int \int \cdots \int h_N(t, v, v_2, v_3, \ldots v_N) \, dv_2 \, dv_3 \ldots dv_N$$

we show, thanks to Theorem 4.2.3, that as $N \to \infty$ we have for all t:

$$\int \int \cdots \int h_N(t, v, v_*, v_3, \ldots v_N) \, dv_3 \, dv_4 \ldots dv_N - h_N^1(t, v) h_N^1(t, v_*) \to 0,$$

and for all t:

$$h_N^1(t, v) \to \hat{f}(t, v).$$

This allows us to confirm that the marginal density h_N^1 of the solution h_N of the principal equation is an approximation to the solution of the Boltzmann equation with nearby constant ρ.

4.2.3   *Interpretation in terms of the Monte-Carlo method*

Theorem 4.2.3 implies that, for fixed t, if $N \to \infty$, the random measure $\overline{\mathbf{V}}_t^N = 1/N \sum_{i=1}^N \delta_{V_{it}^N}$ converges strictly in mean to the measure $\hat{f}(t, v) \, dv$, and therefore also that

$$\frac{\rho}{N} \sum_{i=1}^N \delta_{V_{it}^N} \to f(t, v) \, dv \quad \text{strictly in mean.}$$

Now, a realization of the process $\rho \mathbf{V}_t^N$ corresponds to the following Monte-Carlo method:

- We generate N independent particles V_{i0} distributed following the law h_0. Each particle has weight equal to ρ/N.
- At the end of a time which is an exponential random variable with parameter:

$$\frac{\rho}{N} \sum_{i,j=1}^N q(V_{it} - V_{jt}),$$

  we jump in the following way: with probability

$$\frac{q(V_{it} - V_{jt})}{\sum_{k,l=1}^N q(V_{kt} - V_{lt})}$$

  the pair (i, j) is chosen and the particles i and j change from velocities V_{it}, V_{jt} to velocities V_{it}', V_{jt}' defined by (4.2) for σ equidistributed over the unit sphere.
- We iterate the procedure by choosing, at random, a jump time which is a exponential random variable with parameter:

$$\frac{\rho}{N} \sum_{i,j=1}^N q(V_{it} - V_{jt}),$$

etc.

4.3   Linear and symmetric methods

In this section, we shall present some material complementary to that above proceeding by analogy with what we have done for the transport equation. Indeed, we shall introduce two types of methods: linear and symmetric; symmetric methods are of the type that have been described at the end of the preceding paragraph. (In every case, we define a splitting, at each timestep, between the 'free transport' phase and the 'collision' phase, which introduces an error associated with this time discretization.) The basic principle of the linear method is to linearize the collision operator around the solution at the beginning of the timestep. For symmetric methods, we modify the treatment of the collisions so that they conserve momentum and energy.

Assume therefore that over a timestep $[0, \Delta t]$ the advection phase has been done and let us consider only the collision phase. The space variable no longer plays a part, and we must resolve the following spatially homogeneous equation where f°, is the population at the beginning of the timestep:

$$\frac{\partial f}{\partial t} = Q(f, f), \tag{4.6}$$

$$f(0, v) = f^\circ(v).$$

Since $Q(f, f) = S(f, f) - L(f)f$ we can linearize the operator Q around f° and look for the value at the end of the timestep of the solution f of the equation:

$$\frac{\partial f}{\partial t} + L(f^\circ)\, f = S(f, f^\circ). \tag{4.7}$$

From the properties of Q recalled at the beginning of the chapter, we know that

$$\int [L(f^\circ)f - S(f, f^\circ)]\, dv = 0.$$

Therefore, we deduce conservation of mass:

$$\int f(v)\, dv = \int f^\circ(v)\, dv.$$

But we do not necessarily have:

$$\int [L(f^\circ)f(v) - S(f, f^\circ)(v)]v\, dv = 0.$$

Indeed, given that $f(\Delta t) - f^\circ = 0(\Delta t)$ and that $\int Q(f^\circ, f^\circ)(v)v\, dv = 0$ we see that

$$\left| \int [L(f^\circ)f(\Delta t, v) - S(f(\Delta t), f^\circ)(v)]v\, dv \right| \leq C\Delta t.$$

Therefore, we have:

$$\left| \int f(\Delta t, v) v \, dv - \int f^\circ(v) v \, dv \right| \le C \, \Delta t^2.$$

Likewise, we will have:

$$\left| \int f(\Delta t, v) v^2 \, dv - \int f^\circ(v) v^2 \, dv \right| \le C \, \Delta t^2,$$

that is, the time discretization induces, at each timestep, an error of Δt^2 on the momentum and total energy. We remark that the error accumulated, during a fixed time interval, by the quantities $\int f(t, v) v \, dv$ and $\int f(t, v) |v|^2 \, dv$ tend to 0, as $\Delta t \to 0$. The linearization (4.7) is the basic principle of the method that we shall now describe.

4.3.1 *Short description of the linear Monte-Carlo method*

This method (also called Nambu's method) has been introduced in Nambu (1980) and Illner and Neunzert (1987) .

As for the transport equations, the solution f of equation (4.1) is represented at each moment by a sum of Dirac masses (in space and velocity). Denote by (α_i, x_i, v_i) the weight, the position, and the velocity of the particle numbered i, at the beginning of the timestep. We use an approximation of the solution $f(t, x, v)$ in the form:

$$f(t, x, v) \, dx \, dv \simeq \sum_i \alpha_i \delta_{x_i} \delta_{v_i}$$

- For the advection phase, the particles are displaced in a straight line with the velocities they had at the beginning of the timestep.
- For the collision phase, we proceed locally in each space element M (with volume $\mathrm{Vol}(M)$), denoting by $\overline{f}(t, v)$ the space average of the solution $f(t, x, v)$ in the element. In the timestep under consideration, $\overline{f}$ should satisfy (4.6). Indeed, even if it means making an error of Δt^2, we assume that $\overline{f}$ satisfies the linear equation (4.7).

For a representation of f° at the beginning of the timestep, we must take:

$$f^\circ \, dv \simeq \sum_{x_i \in M} a_i \delta_{v_i} \quad \text{with} \quad a_i = \alpha_i / \mathrm{Vol}(M).$$

For simplicity, we assume from now on that $q(w, \mu)$ does not depend on μ. We then have:

$$L(f^\circ)(v) \simeq \sum_j q(v_j - v) a_j.$$

Following the general principle of Monte-Carlo methods, taking account of the jump operator $S(\cdot, f^\circ)$, each particle i having velocity v_i must jump at a time which is an exponential random variable with parameter $L(f^\circ)(v_i)$. Indeed, even if we must make an error of $(\Delta t)^2$ the particle i jumps at the end of the timestep with probability

$$L(f^\circ)(v_i) = \sum_j q(v_j - v_i) a_j.$$

Providing that:

$$\Delta t \sup_i [L(f^\circ)(v_i)] \leq 1.$$

The law of probability of jumping after this is

$$\pi(A) = S^* 1_A(v_i)/S^* 1(v_i),$$

where S^* denotes the dual operator of $S(\cdot, f^\circ)$ and 1_A the indicator function of the set A. We can verify that

$$S^* 1_A(v_i) = \sum_j q(v_j - v_i) a_j \mathrm{meas}\{\omega \in S^2 / \frac{1}{2}(v_i + v_j) + \frac{1}{2}\omega |v_i - v_j| \in A\}.$$

Numerical algorithm

We first make a copy of all of the particles. For each particle i, the probability that its velocity v_i is modified as

$$p_i = \Delta t \sum_j q(v_j - v_i) \, a_j. \tag{4.8}$$

Then with probability $(q(v_j - v_i) \, a_j)/(\sum_j q(v_j - v_i) \, a_j)$ the jump velocity on the sphere of diameter (v_i, v_j) and the distribution of the new velocity v_i' is uniform over the sphere.

By using the 'fictitious shock' technique we can modify the scheme above. Denote by q_* an estimate of $q(v_i - v_j)$ for the particles of the element; a_* an estimate of the weights of the particles of the element; N the number of particles in the element considered; $a_* = \alpha_*/\mathrm{Vol}(M)$.

Modified numerical algorithm

For each particle, the probability that its velocity v_i is possibly modified during the interval of length Δt is

$$p_* = \Delta t(N - 1)q_* a_*.$$

For each particle i, we choose a partner j chosen uniformly at random from:

$$\{1, 2, \ldots, i, i+1, \ldots N\}.$$

Then, with probability:

$$\frac{q(v_j - v_i)a_j}{q_* a_*},$$

we modify the velocity jumping uniformly on the sphere of diameter $\{v_j, v_i\}$; and with the probability:

$$1 - \frac{q(v_j - v_i)a_j}{q_* a_*},$$

we do not modify the velocity.

For this, the criterion that must be satisfied by Δt (independently of questions of accuracy) is the following:

$$\Delta t(N - 1)q_* \, a_* \leq 1. \tag{4.9}$$

The advantage of the 'fictitious shock' technique is very important from the point of view of the cost of calculation, since we only calculate the quantity $q(v_j - v_i)$ for the pairs of particles (i, j) that are possibly partnered.

Characteristics of the method
- We can use particles of different weights.
- We have conservation of the mass.
- We do not necessarily have conservation of momentum or of the total energy.

If we want precise results, we must make a correction that is delicate and without rigorous justification.

4.3.2 *Description of the symmetric Monte-Carlo methods*

These methods are presented in many references. The most classical Bird method called 'without counting time' (see Bird, 1976), is the best known. See also Babovsky (1986) and Gropengiesser *et al.* (1990).

The basic principle of this type of method is as follows.

Thanks to formula (4.8), we see that the probability for the particle i to collide with the particle j is

$$\Delta t q(v_j - v_i)a_i.$$

Therefore, if all the a_i are equal to a, we see that the probability for i to collide with j is identical to the probability for j to collide with i and becomes:

$$p_{ij} = aq(v_i - v_j)\Delta t.$$

From now on we make the hypothesis within the framework of this section that *all the particles have the same weight a*.

By using the 'fictitious shock' technique, the probability that two particles i and j are subject to a 'potential collision' (in short PC) is given by

$$z = aq_*\Delta t = \alpha q_*\Delta t/\text{Vol}(M).$$

Here, it is once more indispensable to have a sufficiently small timestep, that is, in each element we will have:

$$\Delta t(N - 1)aq_* \leq 1. \tag{4.10}$$

Numerical algorithm. First, we must select a set of pairs of particles (i, j) (with $i = j$) which will be subjected to a PC such that:

(i) a particle cannot be selected more than once,
(ii) the probability that a pair (i, j) is selected is z.

Therefore, the probability that a particle is selected is $(N - 1)z$. Therefore, the expectation of the number of particles selected is $(N(N - 1)/2)z$.

Then for every pair (i, j) of particles we continue in the following way:

- With probability $q(v_i - v_j)/q_*$ the two particles have their velocities modified, they are chosen at random uniformly from a sphere of diameter $\{v_i, v_j\}$. More precisely, the new velocities are given by formulae (4.2) with σ uniformly distributed on S^2.
- With probability $1 - q(v_i - v_j)/q_*$ the particles do not have their velocities modified.

The various methods are distinguished by the method of choosing the pairs (i, j) satisfying (i) and (ii).

In the Neunzert–Babovsky method (Illner and Neunzert 1987), we start by taking $q* = 1/((N - 1)a\Delta t)$ and we make a random permutation of the list $\{1, 2, \ldots, N\}$; then we make pairs of particles whose indices are close after permutation undergo a PC. The disadvantage of this technique is using an estimate of the $q(v_i - v_j)$ and having to test all the chosen pairs for the possibility of a collision.

In Bird's (1976) method, in each element we successively choose p pairs of particles such that

$$2E[p] = zN(N - 1)/2.$$

Then, we carry on as shown above for each pair of particles selected.

Characteristics of symmetric Monte-Carlo methods

- We have conservation of momentum and of global kinetic energy (if we do not take account of the boundary conditions).
- On the other hand, we have a numerical constraint: the particles must all have the same weight.

In the majority of problems we look for a stationary solution, then the solution changes little from one timestep to the next. Specifically, we therefore evaluate an estimate q_* of $q(v_i - v_j)$ in every element in such a way as not to have an estimate q_* which is too large and, as this operation may be relatively expensive, we do not carry it out at each timestep but only occasionally (until the criterion $q(v_i - v_j) \leq q_*$ is violated).

4.4 Implementation of the symmetric methods

In most of the applications, the densities can vary by at least a factor of 20 between different space regions, this is why imposing the restriction that all the particles have the same weight would have serious disadvantages.

If we want to put a significant number of particles in the less dense elements (15 particles often constitute a minimum if we do not want to have too many fluctuations), we are led to have a potentially prohibitory number of particles in the densest elements.

In practice, we assume that the particles have constant weight in zones, which will be called the representative coefficient of the zone. For this we split the domain into zones that have similar physical behaviour, and we continue in the following way:

- In each zone we evaluate a representative by demanding:
  - that it must be a simple ratio of a basic coefficient,
  - that it must be that we have between 20 and 30 particles per element.
- When a particle crosses an interface between two zones, we adjust its weight to the new representative coefficient by a technique of Russian roulette or splitting.

The effectiveness of a Monte-Carlo method for Boltzmann equations will depend on the level of vectorization. For the advection part (which consists of a simple calculation of the trajectory of the particles), it is the same technique as was used for transport, given that the implementation will be simpler as there is no collision during flight. For the collision part it is important not to vectorize over the particles or the collisions in the interior of an element since their number is too small, but first to choose the pairs of particles that collide, then make a global list (over the grid or by zone) and carry out the collision algorithm on this list of pairs.

If we use a parallel machine with distributed memory, a good strategy would be to distribute the processors to subdomains of the mesh. This has been described in Alouges (1993).

4.5 Limits of Monte-Carlo methods

As for the transport equation, there are limits to the effectiveness of Monte-Carlo methods for profound reasons, which are similar to those before. We remark that the medium is very collisional if for a characteristic time Δt we have:

$$Lf(x, \bar{v})\dot{\Delta}t \geq 1.$$

Therefore, an element of diameter Δx will be called 'very collisional' if:

$$\bar{v}Lf(x, \bar{v})^{-1} \leq \Delta x \tag{4.11}$$

(denoting by $\bar{v}$ an average velocity in the element).

We therefore see that, if an element is very collisional, criterion (4.10) cannot be satisfied for a reasonable timestep (of the order of magnitude of $\bar{v}\Delta x$). We conclude that, if we want to be able to implement the Monte-Carlo method effectively, we must adapt the size of the elements to the mean free path so that we are not in the case (4.11).

Physically, this is easy to interpret. Indeed, when the medium is very collisional, we know that the solution of the Boltzmann equation is in fact Maxwellian whose characteristics (density, velocity, and temperature) are solutions of the Euler equations (this is a result which is well known as the Hilbert approximation); for a recent presentation see, for example, Caflisch, 1980. Let us note, however, that, as for the linear transport equations, we can use timesteps that violate criterion (4.10) by adapting the collision technique (see Desvilettes and Peralta, 1995).

4.6   Bibliographic comments

The inherent difficulty in the Boltzmann equation comes from the fact that the collision operator $Q(f, f)$ is completely nonlocal with respect to the velocity variable, but purely local with respect to the spatial variable. This is why the concept of solution for this equation must be the object of great scrutiny. The solutions introduced in Proposition 4.1.3 are called 'renormalized'; the introduction of this concept allows us to integrate properly the equation on the trajectories. (Another tool used in the proof of the existence for the solution of Boltzmann is compactness by averaging, which allows us to obtain compactness for the average velocity $\int f(v)\,dv$ of the solution f.)

Following Kac (1956), Grunbaum (1971), and Sznitman (1984) (and what has been presented in this chapter on the propagation of chaos), the theoretical justification of the approximation of the Boltzmann equation by a principal equation (with dependence with respect to the spatial variable) has been the subject of many published works (cf. Babovsky, 1986; Illner and Neunzert 1987; Ivanov and Rogasinsky, 1988; Wagner, 1992); this approximation, called the Boltzmann–Grad approximation, is the basis of the theory of the BBGKY hierarchy (Bogolyubov, Born, Green, Kirkwood, Yvon), which makes the link between the equations describing the movement of N particles and the Boltzmann type kinetic equations (see Cercignani *et al.*, 1994 for example). Thus, in Graham and Mlard (1995) and Sznitman (1984), there is a probabilistic interpretation of the solution of the equation where the collision operator has been 'regularized' with respect to the space variable (in this case the existence of the solution has been known for a long time, Morgenstern, 1955).

In fact, we have considered only the simplest model of the Boltzmann equations: one type of particles with no internal energy. Often, we must resort to more complex models where we take account of the internal energy of the particles or several types of particles for example. In order to take account of these models, we should first of all write the collision operator correctly (for internal energy, see Bourgat *et al.*, 1994, and for several types of particles, see Sentis and Dellacherie), then generalize what was presented above; see particularly Borgnakke and Larsen (1975), which is the source of numerical simulation of models with internal energy. We shall find in Bird, (1976) many adaptations of the symmetrical methods to various models.

With the Boltzmann equations, we have a nonlinear example where Monte-Carlo methods work very well and have attained great success thanks to their flexibility and their adaptability to various different models, even if the theoretical justification of the methods used is sometimes difficult.

5

The Monte-Carlo method for diffusion equations

In this chapter, we shall first give the probabilistic interpretation of the heat equation due to Brownian motion. This plays a role comparable to characteristic curves that allows us to express the solution of an advection equation of the form:

$$\frac{\partial u}{\partial t} + \frac{\partial}{\partial x}(bu) = 0,$$

b being a function of x. We then show that the results obtained for the Laplacian can be generalized to second-order operators with variable coefficients, the Brownian motion being, in this case, replaced by a diffusion process. We then describe in broad outline the Monte-Carlo algorithms that are used to solve parabolic and elliptic equations numerically.

For these equations, Monte-Carlo methods are only used in preference to the classical finite difference, finite element, or finite volume methods in specific situations. Thus, if we want to solve a problem in a high number of dimensions (e.g. greater than 4), classical methods lead to the inversion of a linear system that is too large to be practicable, and Monte-Carlo methods are often used. Likewise, Monte-Carlo methods are often preferable when we want values of the solution only at certain points of the domain: the case of calculating option prices in finance is typical as we are only interested in a few values of the price. Finally, Monte-Carlo methods are useful when we look for the solution of a degenerate diffusion equation (i.e. when we have a diffusion coefficient that can be zero or when the diffusion operator is a small perturbation of an advection operator).

5.1 Brownian motion and partial differential equations

5.1.1 *Brownian motion*

Definition 5.1.1 We denote by *Brownian motion* a process $(B(t), t \geq 0)$ with real values that:

(i) has continuous trajectories, that is, for almost all ω, the function $t \to B(t, \omega)$ is a continuous function;

(ii) is centred Gaussian; which means that for all integers n, and all $0 \leq t_1 < \cdots < t_n$, the vector $(B(t_1), \dots, B(t_n))$ is a random Gaussian vector with zero expectation;

(iii) has covariance defined by

$$\mathbf{E}[B(t)B(s)] = \inf(t,s), \quad s,t \geq 0.$$

It follows from (iii) that $B(0) = 0$ almost surely and from (ii) and (iii) that Brownian motion has *independent increments*, that is, for all integers n, and $0 < t_1 < t_2 < \cdots < t_n$, the random variables $B(t_1), B(t_2) - B(t_1), \ldots, B(t_n) - B(t_{n-1})$ are independent.

In the same way we define Brownian motion with values in $\mathbf{R}^p$ by replacing (ii) by

$$\mathbf{E}[B(t)B(s)^*] = \inf(t,s)I, \tag{5.1}$$

where $B(t)B(s)^*$ denotes the $p \times p$ matrix with general term $B_i(t)B_j(s)$, and I the identity matrix.

We verify that $(B(t), t \geq 0)$ is a p-dimensional Brownian motion if and only if its coordinates $(B_1(t), t \geq 0)$, $(B_2(t), t \geq 0), \ldots, (B_p(t), t \geq 0)$ are independent scalar Brownian motions.

Remark We can construct Brownian motion as the limit of random walks. Let $(X_n, n \geq 1)$ be a sequence of independent random vectors with dimension p, with common law defined by

$$\mathbf{P}(X_n = \pm\sqrt{p}e_i) = (2p)^{-1}, \quad 1 \leq i \leq p,$$

where $(e_1, \ldots, e_p)$ denotes an orthonormal basis of $\mathbf{R}^p$. We set:

$$S_n = X_1 + \cdots + X_n.$$

The classical central limit theorem tells us that $S_n/\sqrt{n} \to X$ in law, where X is a centred Gaussian vector with variance–covariance matrix I. Donsker's theorem extends this result and confirms that the law of $(S_{[nt]}/\sqrt{n}, t \geq 0)$ converges to that of $(B(t), t \geq 0)$, where B is a Brownian motion of dimension p.

What does this convergence in law mean? We define:

$$X_n(t) = \frac{1}{\sqrt{n}}\sum_{k=0}^{\infty}\left[\left(t - \frac{k}{n}\right)S_{k+1} + \left(\frac{k+1}{n} - t\right)S_k\right]\mathbf{1}_{[\frac{k}{n},\frac{k+1}{n}[}(t)$$

and denote by $\mathbf{P}_n$ the law of $(X_n(t), t \geq 0)$ over the space of continuous functions $C(\mathbf{R}_+; \mathbf{R}^p)$. Then, $\mathbf{P}_n$ converges narrowly (i.e. in the sense of duality with continuous bounded functionals over $C(\mathbf{R}_+; \mathbf{R}^p)$, this set being equipped with the topology of uniform convergence on compact sets) to the probability law of the Brownian motion.

The considerations above show the reader that Brownian motion is a limiting model (or a 'mathematical idealization') for numerous physical phenomena.

With every process $(B(t), t \geq 0)$ with values in $\mathbf{R}^p$, and for all $t \geq 0$, we associate the algebra $\mathcal{F}_t^0 = \sigma(B(s); 0 \leq s \leq t)$, the smallest subalgebra of $\mathcal{A}$ which makes the mappings $\omega \rightarrow B(s, \omega)$ for $0 \leq s \leq t$ measurable. For technical reasons we call the increasing sequence of algebras $(\mathcal{F}_t, t \geq 0)$ *the natural filtration of the process*, $\mathcal{F}_t$ being the algebra generated by the sets $\mathcal{F}_t^0$ and those of zero probability of the algebra $\mathcal{A}$.*

We note that Brownian motion has two mathematical properties that make it particularly interesting. As it has independent increments, it is a *Markov process*. In fact, if $0 < s < t$ and if A is a measurable set of $\mathbf{R}^p$, since Brownian motion has stationary independent increments:

$$\mathbf{P}(B(s+t) - B(s) \in A/\mathcal{F}_s) = \mathbf{P}(B(t) \in A)$$
$$= \frac{1}{(\sqrt{2\pi t})^p} \int_A \exp\left(-\frac{|x|^2}{2t}\right) dx.$$

We then recall the following lemma:

Lemma 5.1.2 *Let X and Y be two random vectors with dimension p, $\mathcal{F}$ a subalgebra of $\mathcal{A}$. If Y is $\mathcal{F}$-measurable and X is independent of $\mathcal{F}$, then for every measurable function ϕ from $\mathbf{R}^{2p}$ to $\mathbf{R}_+$:*

$$\mathbf{E}[\phi(X,Y)/\mathcal{F}] = \int \phi(x,Y)\mathbf{P}_X(dx)$$
$$= \mathbf{E}[\phi(X,Y)/Y].$$

Proof The result is clear when $\phi(x, y)$ is of the form $g(x) \times h(y)$; therefore, being a finite sum of such functions, the general case is deduced by passing to the limit. $\quad\square$

Thanks to Lemma 5.1.2 we can then obtain:

$$\mathbf{P}(B(t) \in A/\mathcal{F}_s) = \mathbf{P}(B(t) - B(s) + B(s) \in A/\mathcal{F}_s)$$
$$= \frac{1}{\sqrt{2\pi(t-s)}} \int_A \exp\left(-\frac{|x - B(s)|^2}{2(t-s)}\right) dx$$
$$= \mathbf{P}(B(t) \in A/B(s)).$$

This is one of the ways to express the fact that B is Markovian.

Further, $(B(t), t \geq 0)$ is a *strong Markov process* in the sense that we shall define. For this we need the idea of a *stopping time*:

Definition 5.1.3 We call a random variable S with values in $\mathbf{R}_+$ a *stopping time* if it satisfies $\{S \leq t\} \in \mathcal{F}_t$ for all $t \geq 0$.

---

*With this definition of $\mathcal{F}_t$, we can confirm that if X and Y are such that $X = Y$ $\mathbf{P}$ almost surely, if X is $\mathcal{F}_t$-measurable then Y is also $\mathcal{F}_t$-measurable. This property is, of course, intuitive and desirable. We assume in what follows that all the filtration algebras considered contain the sets of zero probability.

This condition means that around the trajectory $(B(s), 0 \le s \le t)$ we know whether or not the event $\{S \le t\}$ is realized: we only take the decision to stop at t by observing the trajectory up until the time t.

We call $\mathcal{F}_S$ the algebra of former events at the random moment S, $\mathcal{F}_S = \sigma(B(t \wedge S), t \ge 0)$, the *strong Markov property* can be expressed in the following form:

Proposition 5.1.4 *Let A be a measurable set of $\mathbf{R}^p$, let S be a stopping time and $t \ge 0$:*

$$\mathbf{P}(B(S+t) - B(S) \in A/\mathcal{F}_S) = \mathbf{P}(B(t) \in A).$$

Remark We refer to Karatzas and Shreve (1988) or to Revuz and Yor (1991) for a proof.

We can deduce from Proposition 5.1.4 that the process $(B(S+t) - B(S), t \ge 0)$ has the same law as a Brownian motion starting from 0 and that this Brownian motion is independent of $\mathcal{F}_S$.

As in the case where S is deterministic, we can establish that:

$$\mathbf{P}(B(S+t) \in A/\mathcal{F}_S) = \mathbf{P}(B(S+t) \in A/B(S)).$$

This result extends the Markov property to random times and is called the *strong Markov property*.

The other fundamental property of Brownian motion is to be a *martingale*. We recall, first of all, the definition of a martingale.

Definition 5.1.5 A process $(M(t), t \ge 0)$ is a martingale with respect to a filtration $(\mathcal{F}_t, t \ge 0)$, if, for all $t \ge 0$, $M(t)$ is $\mathcal{F}_t$-measurable and $\mathbf{E}(|M(t)|) < +\infty$, and if we have for all t, s such that $0 \le s \le t$:

$$\mathbf{E}\left(M(t)|\mathcal{F}_s\right) = M(s).$$

It is easy to verify that a Brownian motion is a martingale, in fact, as $B(t) - B(s)$ is independent of $\mathcal{F}_s$ and centred:

$$\mathbf{E}(B(t) - B(s)/\mathcal{F}_s) = \mathbf{E}(B(t) - B(s))$$
$$= 0.$$

This can be rewritten as $\mathbf{E}(B(t)/\mathcal{F}_s) = B(s)$.

5.1.2 *Brownian motion and the heat equation*

In the case of Brownian motion, it is easy to establish Kolmogorov type equations and associated Fokker–Planck equations. From the definition of Brownian motion, the density of the law of $x + B(t)$ is that of a Gaussian law centred at x with variance–covariance matrix tI, which becomes at $y \in \mathbf{R}^p$:

$$G_t(x, y) = (2\pi t)^{-p/2} \exp\left(-|x - y|^2/2t\right).$$

A direct calculation shows that $G_t(x, y)$ satisfies:

$$\frac{\partial G_t}{\partial t}(x, y) = \frac{1}{2}\Delta_y G_t(x, y),$$

$$\frac{\partial G_t}{\partial t}(x, y) = \frac{1}{2}\Delta_x G_t(x, y).$$

Now, if $g \in C_b(\mathbf{R}^p)$, set:

$$u(t, x) = \mathbf{E}\left(g(x + B(t))\right).$$

We therefore have, $u(t, x) = \int_{\mathbf{R}^p} g(y) G_t(x, y)\, dy$. We can then show, justifying the differentiation under the summation sign by Lebesgue's theorem, that:

$$\frac{\partial u(t, x)}{\partial t} - \frac{1}{2}\Delta_x u(t, x) = \int_{\mathbf{R}^p} g(y)\left(\frac{\partial G_t}{\partial t}(x, y) - \frac{1}{2}\Delta_x G_t(x, y)\right) dy = 0.$$

We can then establish that u satisfies a *Kolmogorov equation*:

$$\begin{cases} \dfrac{\partial u}{\partial t}(t, x) = \frac{1}{2}(\Delta u)(t, x), & t > 0, \quad x \in \mathbf{R}^p, \\ u(0, x) = g(x), & x \in \mathbf{R}^p. \end{cases}$$

Now, let us assume that we are interested in the law of $X(t) = X(0) + B(t)$, where $X(0)$ is a random vector with dimension p with law μ_0, independent of $(B(t), t \geq 0)$. It is then clear that for all $t \geq 0$, the law of $X(t)$ has density $p_t(x)$ with respect to the Lebesgue measure and that this density is given by

$$p_t(x) = \int_{\mathbf{R}^p} G_t(x, y)\mu_0(dy).$$

By differentiating under the summation sign, we deduce that the collection of probability densities $\{p_t(\cdot), t > 0\}$ satisfies the *Fokker–Planck equation*:

$$\begin{cases} \dfrac{\partial p_t(x)}{\partial t} = \frac{1}{2}(\Delta p_t)(x), & t > 0, \ x \in \mathbf{R}^p, \\ p_t(\cdot) \to \mu_0, & t \downarrow 0. \end{cases}$$

The link with elliptic problems is more delicate, it follows from the strong Markov property. We shall establish it a little later after having constructed the stochastic integral for Brownian motion.

5.1.3 *The Itô stochastic integral*

We have seen that the trajectories of Brownian motion are continuous. But they are very irregular (this is not surprising, since the increments are independent random variables), in particular, they are neither differentiable nor of finite variation. In fact, Brownian motion has nonzero quadratic variation.

Proposition 5.1.6 *Let $(B(t), t \geq 0)$ be a Brownian motion with values in $\mathbf{R}^p$. For all $t > 0$ and every subdivision $0 = t_0^n < t_1^n < \cdots < t_n^n = t$ whose step $\sup_{i \leq n}(t_i^n - t_{i-1}^n)$ tends to zero as n tends to infinity:*

$$\sum_{i=1}^{n} |B(t_i^n) - B(t_{i-1}^n)|^2 \to p \times t,$$

quadratically in mean, as $n \to \infty$.

Proof We shall restrict ourselves to the case $p = 1$. The random variables $(B(t_i^n) - B(t_{i-1}^n))^2$, $i = 1, 2, \ldots, n$, being independent:

$$\mathrm{Var}\left[\sum_{i=1}^{n}(B(t_i^n) - B(t_{i-1}^n))^2\right] = \sum_{i=1}^{n} \mathrm{Var}\left[(B((t_i^n) - B(t_{i-1}^n))^2\right]$$

$$= 2\sum_{i=1}^{n} \left(t_i^n - t_{i-1}^n\right)^2.$$

This variance therefore tends to 0 as n tends to infinity. Since, further, $\mathbf{E}\left(\sum_{i=1}^{n}(B(t_i^n) - B(t_{i-1}^n))^2\right) = t$ the result is proved. $\quad\Box$

Remark We can prove that a continuous function f with finite variation is necessarily of zero quadratic variation. For this it is sufficient to note that:

$$\sum_{i=1}^{n}(f(t_i^n) - f(t_{i-1}^n))^2 \leq \sup_{i=1,\ldots,n} |f(t_i^n) - f(t_{i-1}^n)| \sum_{i=1}^{n} |f(t_i^n) - f(t_{i-1}^n)|,$$

as the function $(f(s), \leq s \leq t)$ is uniformly continuous the result follows.

Proposition 5.1.6 and the remark above prove that the trajectories of the Brownian motion are almost surely not of finite variation. We see therefore that an integral of the type:

$$\int_0^t \phi(s)\, dB(s)$$

cannot be defined as a Stieltjes integral. However, we can remark that if $(\phi(t), t \geq 0)$ is a deterministic function taking values in $\mathbf{R}$, which is a step function with compact support, of the form:

$$\phi(t) = \sum_{i=1}^{n} \phi_i \mathbf{1}_{]t_{i-1}, t_i]}(t),$$

with n an integer, $0 = t_0 < t_1 < \cdots < t_n$:

$$\mathbf{E}\left(\sum_{i=1}^{n} \phi_i (B(t_i) - B(t_{i-1}))\right) = \sum_{i=1}^{n} \mathbf{E}\left(\phi_i (B(t_i) - B(t_{i-1}))\right)$$

$$= \sum_{i=1}^{n} \phi_i \mathbf{E}\left(B(t_i) - B(t_{i-1})\right)$$

$$= 0.$$

Further,

$$\mathbf{E}\left(\left|\sum_{i=1}^{n} \phi_i (B(t_i) - B(t_{i-1}))\right|^2\right) = \mathrm{Var}\left(\sum_{i=1}^{n} \phi_i (B(t_i) - B(t_{i-1}))\right)$$

$$= \sum_{i=1}^{n} \phi_i^2 (t_{i+i} - t_i)$$

$$= \int_0^\infty \phi(t)^2 \, dt.$$

It is then easy to see that the mapping:

$$\phi \to \int_0^\infty \phi(t) \, dB(t),$$

defined for ϕ a step function of the form above by

$$\int_0^\infty \phi(t) \, dB(t) := \sum_{i=1}^{n} \phi_i (B(t_i) - B(t_{i-1}))$$

extends to an isometric linear mapping from $L^2(\mathbf{R}_+)$ to $L^2(\Omega)$. We define the *Wiener integral* of a deterministic function from $L^2(\mathbf{R}_+)$ with respect to the Brownian motion. We shall see that we can construct an integral for a large class of random processes. This integral is called the Itô integral and we shall give the broad outline of its construction. We refer to Karatzas and Shreve (1988) for a detailed construction.

Construction of the Itô integral. The definition of the *Itô integral* of a nonanticipative random process $(\phi(t), t \geq 0)$ is analogous to that of the

Wiener integral. Technically, we shall assume that the process ϕ is *adapted* to the filtration $(\mathcal{F}_t, t \geq 0)$,[†] that is, for all $t \geq 0$, ϕ_t is $\mathcal{F}_t$-measurable. We also say that the process ϕ is *nonanticipative* (it does not anticipate the future increments of the Brownian motion).

We assume, initially, that the process ϕ takes values in $\mathbf{R}$ and that $(B(t), t \geq 0)$ is a real Brownian motion. We shall denote by $\mathbf{M}^2$:

$$\mathbf{M}^2 = \left\{ (\phi(s), s \geq 0) \text{ adapted to } \mathcal{F}_t \text{ and such that} \right.$$

$$\left. \times \mathbf{E}\left(\int_0^{+\infty} |\phi(s)|^2 \, ds \right) < +\infty \right\}.$$

The calculations made above with ϕ being deterministic extend to the case $\phi \in \mathbf{M}^2$. For this, we say that a process ϕ is an *elementary process* if:

$$\phi(t, \omega) = \sum_{i=1}^{n} \phi_i(\omega) \mathbf{1}_{]t_{i-1}, t_i]}(t),$$

where $0 = t_0 < t_1 < \cdots < t_n$ and the random variables ϕ_i are $\mathcal{F}_{t_{i-1}}$-measurable and bounded. We start by defining the stochastic integral of an elementary process ϕ by

$$\int_0^{+\infty} \phi(s) \, dB(s) = \sum_{i=1}^{n} \phi_i \left(B(t_i) - B(t_{i-1}) \right).$$

We can also prove that:

$$\mathbf{E}\left(\int_0^{+\infty} \phi(s) \, dB(s) \right) = \mathbf{E}\left(\sum_{i=1}^{n} \phi_i \left(B(t_i) - B(t_{i-1}) \right) \right)$$

$$= \mathbf{E}\left(\sum_{i=1}^{n} \phi_i \mathbf{E}\left(B(t_i) - B(t_{i-1}) | \mathcal{F}_{t_{i-1}} \right) \right).$$

Since Brownian motion has centred independent increments, we have:

$$\mathbf{E}\left(B(t_i) - B(t_{i-1}) | \mathcal{F}_{t_{i-1}} \right) = \mathbf{E}\left(B(t_i) - B(t_{i-1}) \right) = 0,$$

and therefore $\mathbf{E}\left(\int_0^T \phi(s) \, dB(s) \right) = 0$. Further, we note that if $\phi = \sum_{i=1}^{n} \phi_i \mathbf{1}_{]t_{i-1}, t_i]}$ is a elementary process we have, for $i < j$:

$$\mathbf{E}\left[\phi_i(B(t_i) - B(t_{i-1}))\phi_j(B(t_j) - B(t_{j-1})) \right] = 0,$$

[†]We can replace the filtration $(\mathcal{F}_t, t \geq 0)$ by a finer one. That is such that, for all $t \geq 0$, $\mathcal{F}_t \subset \mathcal{G}_t$. $(\mathcal{G}_t, t \geq 0)$, as long as for all t, $(B(t+s) - B(t), s \geq 0)$ is independent of $\mathcal{G}_t$. We say in this case that $(B(t), t \geq 0)$ is a $\mathcal{G}_t$-Brownian motion. It is this property that is crucial in the construction of the stochastic integral that follows. We therefore assume from now on that B is an $\mathcal{F}_t$-Brownian motion.

since the random variable $B(t_j) - B(t_{j-1})$ is centred and independent of $\phi_i(B(t_i) - B(t_{i-1}))\phi_j$, which is $\mathcal{F}_{t_{j-1}}$ measurable. This allows us to prove, for an elementary process, that:

$$
\begin{aligned}
\mathbf{E}\left(\left|\int_0^\infty \phi(t)\,dB(t)\right|^2\right) &= \sum_{i=1}^n \sum_{j=1}^n \mathbf{E}\left(\phi_i\phi_j\left(B(t_i) - B(t_{i-1})\right)\right. \\
&\qquad\qquad \left. \times \left(B(t_j) - B(t_{j-1})\right)\right) \\
&= \sum_{i=1}^n \mathbf{E}\left(\phi_i^2\left(B(t_i) - B(t_{i-1})\right)^2\right) \\
&= \sum_{i=1}^n \mathbf{E}\left(\phi_i^2\mathbf{E}\left(B(t_i) - B(t_{i-1})\right)^2\,|\mathcal{F}_{t_{j-1}}\right) \\
&= \sum_{i=1}^n \mathbf{E}\left(\phi_i^2(t_i - t_{i-1})\right) = \mathbf{E}\left(\int_0^\infty \phi(t)^2\,dt\right).
\end{aligned}
$$

This establishes an isometry property that allows us to extend by density the stochastic integral of elementary processes. For this, we use the following property, of which there is a proof in Karatzas and Schreve (1988, p. 134, problem 2.5): for every process $\phi \in \mathbf{M}^2$ there exists a sequence of elementary processes $(\phi_n(s), 0s \geq 0)$ such that:

$$
\lim_{n \to +\infty} \mathbf{E}\left(\int_0^{+\infty} |\phi(s) - \phi_n(s)|^2\,ds\right) = 0.
$$

This allows us to prolong the stochastic integral of elementary processes to all the elements of $\mathbf{M}^2$. We continue to denote this stochastic integral for an element of $\mathbf{M}^2$:

$$
\int_0^{+\infty} \phi(s)\,dB(s).
$$

By passing to the limit in the preceding equalities we can prove that this mapping is linear and that for every process ϕ of $\mathbf{M}^2$, we have:

$$
\mathbf{E}\left(\int_0^\infty \phi(s)\,dB(s)\right) = 0
$$

and

$$
\mathbf{E}\left(\left|\int_0^\infty \phi(s)\,dB(s)\right|^2\right) = \mathbf{E}\left(\int_0^\infty |\phi(s)|^2\,ds\right).
$$

We can then define, for all $t \geq 0$:

$$
\int_0^t \phi(s)\,dB(s) := \int_0^\infty \phi(s)\mathbf{1}_{[0,t]}(s)\,dB(s),
$$

for every process $\phi \in \bigcap_{T \geq 0} \mathbf{M}_T^2$, with:

$$\mathbf{M}_T^2 = \left\{ (\phi(s), 0 \leq s \leq T) \text{ adapted and such that} \right.$$

$$\times \mathbf{E} \left(\int_0^T |\phi(s)|^2 \, ds \right) < +\infty \right\}.$$

In fact, we can likewise define the Itô stochastic integral:

$$\int_0^t \phi(s) \, dB(s),$$

with $\phi \in \bigcap_{T \geq 0} \mathbf{M}_{T,loc}^2$, where $\mathbf{M}_{T,loc}^2$ is the set of adapted processes that satisfy:

$$\int_0^T |\phi(s)|^2 \, ds < \infty \quad \text{almost surely.}$$

Note that, in every case, we can construct the stochastic integral:

$$\left(\int_0^t \phi(s) \, dB(s), t \geq 0 \right),$$

such that it is almost surely continuous in t and that if $\phi \in \mathbf{M}_T^2$, then for $t \leq T$:

$$\mathbf{E} \left(\int_0^t \phi(s) \, dB(s) \right) = 0,$$

$$\mathbf{E} \left(\left(\int_0^t \phi(s) \, dB(s) \right)^2 \right) = \mathbf{E} \left(\int_0^t |\phi(s)|^2 \, ds \right).$$

In the case $\phi \in \mathbf{M}_{T,loc}^2$, we can only assert that:

$$\mathbf{E} \left(\left(\int_0^t \phi(s) \, dB(s) \right)^2 \right) \leq \mathbf{E} \left(\int_0^t |\phi(s)|^2 \, ds \right).$$

Remark If $(B(t), t \geq 0)$ is a p-dimensional Brownian motion, and $\phi \in (\mathbf{M}_{T,loc}^2)^{n \times p}$, then we define the n-dimensional process:

$$\left(\int_0^t \phi(s) \, dB(s), 0 \leq t \leq T \right)$$

by

$$\left(\int_0^t \phi(s) \, dB(s) \right)_i = \sum_{j=1}^n \int_0^t \phi_{ij}(s) \, dB_j(s), \ 1 \leq i \leq p.$$

Stochastic integral and martingale. We have seen that Brownian motion is a martingale. The idea of a stochastic integral allows us to construct, starting from Brownian motion, numerous martingales. We shall see that this idea is very useful to establish links between stochastic differential equations and partial differential equations. The following result shows that the process $t \to \int_0^t \phi(s)\,dB(s)$ is a martingale.

Proposition 5.1.7 *Let $(\phi(t), t \geq 0)$ be a process belonging to $\mathbf{M}_T^2$. Then the process $(\int_0^t \phi(s)\,dB(s)), 0 \leq t \leq T)$ is a martingale with respect to the filtration $(\mathcal{F}_t, 0 \leq t \leq T)$. In particular, if τ is a stopping time with respect to $(\mathcal{F}_t, 0 \leq t \leq T)$ smaller than T, we have:*

$$\mathbf{E}\left(\int_0^\tau \phi(s)\,dB(s)\right) = 0.$$

Remark The last assertion of this proposition is a particular case of the *stopping theorem* for the martingale $\left(\int_0^t \phi(s)\,dB(s), 0 \leq t \leq T\right)$. This result plays an important role when we try to establish probabilistic interpretations for some elliptic equations.

Proof We have seen that if ϕ is a process of $\mathbf{M}_T^2$, we have:

$$\mathbf{E}\left(\int_0^T \phi(s)\,dB(s)\right) = 0.$$

We then remark that, if ϕ is a process of $\mathbf{M}_T^2$ and τ a stopping time bounded by T, the process $(\mathbf{1}_{\{t<\tau\}}\phi(t), t \geq 0)$ is adapted to $\mathbf{M}_T^2$. We can, further, show the following intuitive (but not completely obvious) result a priori, almost surely:

$$\int_0^T \mathbf{1}_{\{s<\tau\}}\phi(s)\,dB(s) = \left(\int_0^t \phi(s)\,dB(s)\right)_{t=\tau}.$$

It is then easy to show that:

$$\mathbf{E}\left(\int_0^\tau \phi(s)\,dB(s)\right) = \mathbf{E}\left(\int_0^T \mathbf{1}_{\{t<\tau\}}(s)\phi(s)\,dB(s)\right) = 0.$$

We can then deduce the martingale property of the process $M(t) = \int_0^t \phi(s)\,dB(s)$ by remarking that, if A is an $\mathcal{F}_s$-measurable event, $\tau = s\mathbf{1}_A + t\mathbf{1}_{A^c}$ is a stopping time. From which:

$$0 = \mathbf{E}\left(M(\tau)\right) = \mathbf{E}\left(M(s)\mathbf{1}_A\right) + \mathbf{E}\left(M(t)\right) - \mathbf{E}\left(M(t)\mathbf{1}_A\right).$$

Since $\mathbf{E}\left(M(t)\right) = 0$, we therefore deduce that $\mathbf{E}\left(M(s)\mathbf{1}_A\right) = \mathbf{E}\left(M(t)\mathbf{1}_A\right)$. This allows us to conclude the result. $\square$

The Itô formula.   The essential element that allows us to carry out
the calculations that occur in stochastic integrals is the *Itô formula*. We
shall restrict ourselves to stating the result and indicating a sketch of the
proof in the case of Brownian motion. For a complete proof we refer to
Karatzas and Shreve (1988) or to Revuz and Yor (1991).

We start by stating the Itô formula for Brownian motion.

Proposition 5.1.8 *Let $(B(t), t \geq 0)$ be a Brownian motion of dimension
p. Let Φ be a function of class $C^2(\mathbf{R}^p)$, then:*

$$\Phi(B(t)) = \Phi(B(0)) + \sum_{i=1}^{p} \int_0^t \Phi_i'(B(s)) \, dB_i(s)$$

$$+ \frac{1}{2} \sum_{i=1}^{p} \int_0^t \Phi_{ii}''(B(s)) \, ds.$$

Proof   We shall sketch the proof in the case $p = 1$. Set $t_i^n = (i/n)\, t$. With
the help of Taylor's formula, we obtain:

$$\Phi(B(t_{i+1}^n)) - \Phi(B(t_i^n)) = \Phi'(B(t_i^n))(B(t_{i+1}^n) - B(t_i^n))$$

$$+ \frac{1}{2} \Phi''(\theta_i^n)(B(t_{i+1}^n) - B(t_i^n))^2,$$

with θ_i^n a point between $B(t_i^n)$ and $B(t_{i+1}^n)$. It is not difficult to show that:

$$\sum_{i=0}^{n-1} \Phi'(B(t_i^n)) \left(B(t_{i+1}^n) - B(t_i^n)\right) \text{ tends to } \int_0^t \Phi'(B(s)) \, dB(s),$$

in probability as $n \to \infty$. It follows from Proposition 5.4.1 and from the
continuity of $t \to \Phi''(B(t))$, with the help of an argument from integration
theory, that:

$$\sum_{i=0}^{n-1} \Phi''(\theta_i^n)(B(t_{i+1}^n) - B(t_i^n))^2 \quad \text{tends, in probability, to } \int_0^t \Phi''(B(s)) \, ds,$$

as n tends to ∞.   □

In fact, we can show in an analogous way a more general Itô formula.
We call an *Itô process* a process of the form:

$$X(t) = X(0) + \int_0^t \psi(s) \, ds + \int_0^t \phi(s) \, dB(s), \quad t \geq 0, \tag{5.2}$$

with $X(0)$ a random vector of dimension n independent of $(B(t), t \geq 0)$,
$\psi \in \cap_T (\mathbf{M}_{T,loc}^2)^n$, $\phi \in \cap_T (\mathbf{M}_{T,loc}^2)^{n \times p}$, and $(B(t), t \geq 0)$ p-dimensional
Brownian motion.

Proposition 5.1.9 *Let $(X(t), t \geq 0)$ be an Itô process of the form (5.2), and $\Phi \in C^2(\mathbf{R}^n)$. Then:*

$$
\Phi(X(t)) = \Phi(X(0)) + \sum_{i=1}^{n} \int_0^t \Phi_i'(X(s)) \, dX_i(s)
$$
$$
+ \frac{1}{2} \sum_{i,j=1}^{n} \int_0^t \Phi_{ij}''(X(s)) \, d\langle X_i, X_j \rangle_s,
\tag{5.3}
$$

with:

- $dX_i(s) = \psi_i(s) \, ds + \sum_{j=1}^{p} \phi_{ij}(s) \, dB_j(s)$,
- $d\langle X_i, X_j \rangle_s = \sum_{k=1}^{p} \phi_{ik}(s) \phi_{jk}(s) \, ds$.

We most often write the Itô formula in a more convenient differential form:

$$
d\Phi(X(t)) = \sum_{i=1}^{n} \Phi_i'(X(s)) dX_i(s) + \frac{1}{2} \sum_{i,j=1}^{n} \Phi_{ij}''(X(s)) \, d\langle X_i, X_j \rangle_s.
$$

Remark We note the particular form this formula takes when $X_1(t) = t$. We denote $\bar{X}(t) = (X_2(t), \ldots, X_n(t))$. By applying the preceding formula and by remarking that $\langle X_1, X_i \rangle_t = 0$, for $i = 1, \ldots, n$, we obtain:

$$
d\Phi(t, \bar{X}(t)) = \Phi(0, \bar{X}(0)) + \frac{\partial \Phi}{\partial t}(s, \bar{X}(s)) \, ds + \sum_{i=2}^{n} \Phi_i'(s, \bar{X}(s)) \, dX_i(s)
$$
$$
+ \frac{1}{2} \sum_{i,j=2}^{n} \Phi_{ij}''(s, \bar{X}(s)) \, d\langle X_i, X_j \rangle_s.
$$

$$\tag{5.4}$$

In particular, if $(X(t), t \geq 0)$ is an Itô process with values in $\mathbf{R}$, if c is a real constant and $Y(t) = X(t)e^{-ct}$, noting that $Y(t) = f(t, X_t)$ with $f(t, x) = xe^{-ct}$ we obtain:

$$
dY(t) = -cX(t)e^{-ct} \, dt + e^{-ct} \, dX(t).
\tag{5.5}
$$

This remark will be useful to us in what follows.

We shall show how the Itô formula allows us to establish a link between Brownian motion and some elliptic equations.

5.1.4 *Brownian motion and the Dirichlet problem*

Let D be a regular bounded domain of $\mathbf{R}^p$. We define the exit time from D starting from x:

$$
S^x = \inf\{t \geq 0; x + B(t) \in D^c\}.
$$

We can show that $x + B(t)$ leaves D almost surely in finite time, and that S^x is a stopping time. By continuity of the trajectories, we see that

$x + B(S^x)$ takes its values on the boundary of D, which we denote ∂D. Given f continuous and bounded on ∂D, we set:

$$u(x) = \mathbf{E}[f(x + B(S^x))].$$

The following result shows that the function u defined in this way is the solution of a Dirichlet problem.

Theorem 5.1.10 *Assume that the boundary ∂D of an open bounded set D is regular and that f is continuous and bounded on ∂D. Then, $u(x) = \mathbf{E}[f(x+B(S^x))]$ is the unique solution of class $C^2(D)\cap C(\bar{D})$ of the Dirichlet problem:*

$$\begin{cases} \Delta u(x) = 0, & x \in D, \\ u(x) = f(x), & x \in \partial D. \end{cases}$$

Proof The first equation means that u is harmonic in D. We can show that this property is equivalent to saying that, for all $x \in D$, $r > 0$ such that $B(x,r) = \{x \in \mathbf{R}^p, |x| \leq r\} \subset D$,

$$u(x) = \int_{\{|y-x|=r\}} u(y)\sigma_{x,r}(dy),$$

where $\sigma_{x,r}$ denotes the uniform measure over the sphere $B(x,r)$ comprised of points at a distance r from x. We shall verify that the function defined by $u(x) = \mathbf{E}[f(x + B(S^x))]$ satisfies this last property. For this, we denote:

$$T_r^x = \inf\{t \geq 0; X^x(t) \in B(x,r)^c\},$$

where $X^x(t) = x + B(t)$. We then have:

$$u(x) = \mathbf{E}\left(\mathbf{E}\left(f(X^x(S^x))/X^x(T_r^x)\right)\right).$$

The fact that T_r^x is a stopping time then allows us to show (see Proposition 5.1.4) that:

$$(X^x(s + T_r^x) - X^x(T_r^x), s \geq 0)$$

is a Brownian motion starting from 0 at time 0 independent of the random variable $X^x(T_r^x)$. Since, for r sufficiently small, we have $T_r^x < S^x$, we therefore obtain:

$$\mathbf{E}\left(f(X^x(S^x))/X^x(T_r^x)\right) = u\left(X^x(T_r^x)\right).$$

Further, the invariance of Brownian motion under rotation allows us to prove that the law of $X^x(T_r^x)$ is uniform over the sphere $B(x,r)$. From this we deduce that, for small r:

$$u(x) = \int_{\{|y-x|=r\}} u(y)\sigma_{x,r}(dy),$$

u is therefore a harmonic function over D.

The fact that if $x \in \partial D$ we have $\lim_{y \to x, y \in D} u(y) = f(x)$ can be understood intuitively if we remark that, for $x \in \partial D$, $T_r^x = 0$. However, a precise proof is delicate and relies on the regularity hypotheses of the open set D and of the function f that we have not made explicit here. We refer to Rogers and Williams (1994) for details and a proof of this result.

To finish the proof we have to show the uniqueness of the solution of the Dirichlet problem. This can be done by using a classical argument resulting from the maximum principal. $\square$

Remark We can give another proof of the uniqueness in the class of functions $C^2(D) \cap C(\bar{D})$ by using the Itô formula. This proof can be adapted to more general cases and allows us formally to recover the link between Brownian motion and a solution to the Dirichlet problem. For reasons of simplicity, we give the idea of the proof in the case $p = 1$ (the general case is treated almost identically). For this note that, if u is of class C^2, by using the Itô formula:

$$u(x + B(t)) = u(x) + \frac{1}{2}\int_0^t u''(x + B(s))\, ds + \int_0^t u'(x + B(s))\, dB(s).$$

We deduce that:

$$u(x + B(S^x)) = u(x) + \frac{1}{2}\int_0^{S^x} u''(x + B(s))\, ds + \int_0^{S^x} u'(x + B(s))\, dB(s).$$

Now if u is a solution of class C^2, satisfying $u'' = 0$ in Ω we have:

$$u(x + B(S^x)) = u(x) + \int_0^{S^x} u'(x + B(s))\, dB(s).$$

But since S^x is the exit time from a bounded domain we can prove that $\mathbf{E}(S^x) < +\infty$. This allows us to obtain (approximating S^x by a family of bounded stopping time $(\inf(S^x, n)$, for integer $n)$ that:

$$\mathbf{E}\left(\int_0^{S^x} u'(x + B(s))\, dB(s)\right) = 0.$$

We deduce that $u(x) = \mathbf{E}\left(u(x + B(S^x))\right) = \mathbf{E}\left(f(x + B(S^x))\right)$.

5.1.5   Feynman–Kac formula

We can also, with the help of the Itô formula, give a probabilistic representation formula for the unique solution of the following parabolic equation:

$$\begin{cases} \dfrac{\partial u}{\partial t}(t, x) = \dfrac{1}{2}\Delta u(t, x) + c(x)u(t, x) + f(x), & t > 0, \ x \in \mathbf{R}^p, \\ u(0, x) = g(x), \end{cases}$$

with $c, f, g \in C_b(\mathbf{R}^p)$. We then have the following proposition:

Proposition 5.1.11

$$
u(t, x) = \mathbf{E} \left[g(x + B(t)) \exp \left(\int_0^t c(x + B(s)) \, ds \right) \right.
$$
$$
\left. + \int_0^t f(x + B(s)) \exp \left(\int_0^s c(x + B(r)) \, dr \right) \, ds \right]
$$

Proof We shall carry out the proof under the supplementary hypothesis that $u \in C_b^{1,2}(\mathbf{R}_+ \times \mathbf{R}^p)$. Set $v(s, x) = u(t - s, x)$, $0 \le s \le t$, and let us consider the process:

$$
v(s, x + B(s)) \exp \left(\int_0^s c(x + B(r)) \, dr \right).
$$

By applying the Itô formula to the function f with $f(x, y) = v(t, x) \exp(y)$ and to the pair of the process $\left(x + B(t), \int_0^t c(x + B(s)) \, ds \right)$, we obtain:

$$
v(t, x + B(t)) \exp \left(\int_0^t c(x + B(s)) \, ds \right)
$$
$$
= v(0, x) + \int_0^t \left(\frac{\partial v}{\partial t} + \frac{1}{2} \Delta v + cv \right) (s, x + B(s)) \exp \left(\int_0^s c(x + B(r)) \, dr \right) \, ds
$$
$$
+ \sum_{i=1}^p \int_0^t \frac{\partial v}{\partial x_i} (t, x + B(s)) \exp \left(\int_0^s c(x + B(r)) \, dr \right) \, dB_i(s).
$$

Since u satisfies the equation $\partial u / \partial t = \frac{1}{2} \Delta u + cu + f$, by taking the expected value, we obtain the stated formula. ∎

Remark Note that we obtain in this proposition a result that is very close to that of Theorem 2.3.6, the only difference being the nature of the process taken to represent the partial differential equation: a transport process in the first case and a diffusion process in the second.

We have an analogous formula for an equation in a domain $D \subset \mathbf{R}^p$, with Dirichlet boundary condition, in terms of the 'stopped at the boundary ∂D' process (cf. Section 5.2). When the equation has Neumann boundary conditions, it is interpreted as a 'reflection on the boundary of D process'. We refer the reader to Bensoussan and Lions, (1978, 1982) and Dautray (1989) for further information.

5.2   Probabilistic representations and the diffusion process

We shall see that we can represent the solutions of the Dirichlet problem ($\Delta u = 0$ in D and $u = f$ over ∂D) and parabolic equations of the type of the heat equation by using Brownian motion. The solutions of these equations are then in the form of an expected value of a random variable which lends itself to a Monte-Carlo method. We shall extend this type of formula to particular second-order linear operators and to nearby problems. To express these representation formulae we need to introduce a new class of processes that generalize Brownian motion: diffusion processes. It is the trajectories of these processes that play the role of 'random characteristic curves'. We shall construct these processes as solutions of a stochastic differential equation and we shall show the natural link that connects them to second-order linear operators.

5.2.1   *Stochastic differential equations*

A stochastic differential equation is a generalization of the idea of an ordinary differential equation:

$$\dot{X}(t) = b(X(t)), \quad X(0) = x.$$

We perturb this equation by the derivative of a Brownian motion $B(t)$ (multiplied by a diffusion coefficient $\sigma(x)$):

$$\dot{X}(t) = b(X(t)) + \sigma(X(t))\dot{B}(t).$$

This statement does not have strict meaning, since Brownian motion is not differentiable, and we prefer to write this equation in an integral form:

$$X(t) = x + \int_0^t b(X(s))\,ds + \int_0^t \sigma(X(s))\,dB(s). \qquad (5.6)$$

We often use (5.6) in the symbolic form:

$$\begin{cases} dX(t) = b\,(X(t))\,dt + \sigma\,(X(t))\,dB(t), \\ X(0) = x. \end{cases}$$

We call the equations 'stochastic differential equations'. The solution $(X(t), t \geq 0)$ of Equation (5.6) is called a 'diffusion process' (or by abuse of language, a 'diffusion').

Let us make the notation precise. We are given $b : \mathbf{R} \to \mathbf{R}$, $\sigma : \mathbf{R} \to \mathbf{R}$, and $(B(t), t \geq 0)$ a Brownian motion with respect to the filtration $(\mathcal{F}_t, t \geq 0)$. Finding a solution to Equation (5.6) means finding a continuous stochastic process $(X(t), t \geq 0)$, such that for all t, $X(t)$ is $\mathcal{F}_t$ measurable and which satisfies:

- for all $t \geq 0$, the integrals $\int_0^t b(X(s))\, ds$ and $\int_0^t \sigma(X(s))\, dB(s)$ have a sense:

$$\int_0^t |b(X(s))|\, ds < +\infty \quad \text{and} \quad \int_0^t |\sigma(X(s))|^2\, ds < +\infty \quad \text{almost surely:}$$

- for all $t \geq 0$:

$$\text{almost surely} \quad X(t) = x + \int_0^t b\,(X(s))\, ds + \int_0^t \sigma\,(X(s))\, dB(s).$$

Theorem 5.2.1 gives sufficient conditions on b and σ to have an existence and uniqueness result for (5.6).

Theorem 5.2.1 *If b and σ are functions such that there exists $K < +\infty$, with $|b(x) - b(y)| + |\sigma(x) - \sigma(y)| \leq K|x - y|$, then, for all $T \geq 0$, (5.6) has a unique solution in the interval $[0, T]$. Further this solution satisfies:*

$$\mathbf{E}\left(\sup_{0 \leq t \leq T} |X(s)|^2 \right) < +\infty.$$

Remark The uniqueness means that if $(X(t))_{0 \leq t \leq T}$ and $(Y(t))_{0 \leq t \leq T}$ are two solutions of (5.6), then almost surely $\forall 0 \leq t \leq T$, $X(t) = Y(t)$. For a proof we refer to Karatzas and Shreve (1988).

We shall give several examples of classical diffusion processes.

Example 4 The Ornstein–Uhlenbeck process is the unique solution of the following equation:

$$\begin{cases} dX(t) = -cX(t)\, dt + \sigma\, dB(t), \\ X(0) = x. \end{cases}$$

We can make this solution explicit. In effect, we set $Y(t) = X(t)e^{ct}$, by using Equation (5.5) we obtain:

$$dY(t) = dX(t)e^{ct} + X(t)\, d(e^{ct}).$$

We deduce that $dY(t) = \sigma e^{ct}\, dB(t)$, then that:

$$X(t) = xe^{-ct} + \sigma e^{-ct} \int_0^t e^{cs}\, dB(s).$$

Example 5 The Ornstein–Uhlenbeck process is a particular case of the following problem. We are given a function ϕ from $\mathbf{R}^n$ to $\mathbf{R}$ which will play the role of a potential. We shall consider the diffusion having the constant diffusion coefficient and with derivative:

$$b(x) = -\nabla\phi(x).$$

We are then interested in the unique solution process (subject to hypotheses on ϕ) of the equation:

$$dX(t) = -\nabla\phi(X(t))\,dt + \sigma\,dB(t)$$

$(B(t), t \geq 0)$ being a Brownian motion and σ being a given real number.

The Ornstein–Uhlenbeck process corresponds to the following choice for ϕ:

$$\phi(x) = \frac{c}{2}|x|^2.$$

We can envisage complicating the potential by setting, for example:

$$\phi(x) = \frac{c}{2}|x|^2 - \epsilon|x|^4.$$

Example 6 In the examples of statistical mechanics, we are led to consider that the particle, or more generally the mechanical system, is governed by a stochastic differential equation of the type:

$$\begin{cases} dX(t) = V(t)\,dt, \\ dV(t) = b(t, X(t), V(t))\,dt + \sigma(t, X(t), V(t))\,dB(t), \end{cases}$$

X representing the position of the system and V its velocity. This equation is close to the random evolutions that were studied in Chapter 2: the dynamics describing the evolution of the velocity is described here using a diffusion and not a jump process as in Chapter 2.

Example 7 The most useful process for financial modelling is the *Black–Scholes model*. We assume that $(S(t), t \geq 0)$ is the unique solution of the equation:

$$dS(t) = S(t)\,(r\,dt + \sigma\,dB(t)),$$

where r and σ are two real numbers and $(B(t), t \geq 0)$ is a Brownian motion.

By applying the Itô formula to $\log(S(t))$, we see that:

$$d\log(S(t)) = \left(r - \frac{\sigma^2}{2}\right)dt + \sigma\,dB(t).$$

The random variable $S(t)$ is therefore written in the form:

$$S(t) = x\exp\left(\left(r - \frac{\sigma^2}{2}\right)t + \sigma B(t)\right).$$

This process is also often called a 'geometric Brownian motion' since it has independent multiplicative increments. We note further that the random variable $S(t)$ follows a log-normal law (that is the Gaussian exponential law).

We can generalize the definition of stochastic differential equations to the case where the process evolves in $\mathbf{R}^n$. We are given:

- $B = (B^1, \ldots, B^p)$ a p-dimensional Brownian motion with respect to a filtration $(\mathcal{F}_t, t \geq 0)$;
- $b : \mathbf{R}^n \to \mathbf{R}^n$, $b(x) = (b^1(x), \ldots, b^n(x))$;
- $\sigma : \mathbf{R}^n \to \mathbf{R}^{n \times p}$ (the set of $n \times p$ matrices),

$$\sigma(x) = (\sigma_{i,j}(x))_{1 \leq i \leq n, 1 \leq j \leq p},$$

and we consider the stochastic differential equation:

$$X(t) = x + \int_0^t b(X(s))\, ds + \int_0^t \sigma(X(s))\, dB(s), \qquad (5.7)$$

where we must understand that we are looking for a process $(X(t), t \geq 0)$ with values in $\mathbf{R}^n$ adapted to $(\mathcal{F}_t, t \geq 0)$ and such that, for all t and for all $1 \leq i \leq n$, we have almost surely:

$$X_i(t) = x_i + \int_0^t b_i(X(s))\, ds + \sum_{j=1}^p \int_0^t \sigma_{ij}(X(s))\, dB^j(s).$$

The existence and uniqueness theorem generalizes in the following way:

Theorem 5.2.2 *We assume that* $|b(x) - b(y)| + |\sigma(x) - \sigma(y)| \leq K|x - y|$, *then there exists a unique solution to Equation (5.7).*

Let us give an example of multidimensional diffusion.

Example 8 We can generalize the Black–Scholes model using several assets. A natural way to proceed is as follows, we assume that $(S_1(t), \ldots, S_N(t), t \geq 0)$ are solutions of:

$$\begin{cases} dS_1(t) = S_1(t)\left(r\, dt + \sum_{j=1,N} \sigma_{1j}\, dB_j(t)\right), & S_1(0) = x_1, \\ \cdots \qquad \cdots \\ dS_N(t) = S_N(t)\left(r\, dt + \sum_{j=1,N} \sigma_{Nj}\, dB_j(t)\right), & S_N(0) = x_N, \end{cases}$$

where $(B(t), t \geq 0)$ is an N-dimensional Brownian motion, r a positive real number, and σ a given matrix.

We can verify that each of the coordinates $S_i(t)$ follow an identical model, in law, to that of Black–Scholes described in Example 7. It is enough, for this, to verify that $(S_i(t), t \geq 0)$ is the solution of:

$$dS_i(t) = S_i(t)\left(r\, dt + \sigma_i\, d\tilde{B}_i\right), \quad S_i(0) = x_i,$$

with:

$$\sigma_i^2 = \sum_{j=1}^{N} \sigma_{ij}^2 \quad \text{and} \quad \tilde{B}_i(t) = \frac{\sum_{j=1,N} \sigma_{ij} B_j(t)}{\sigma_i},$$

and to remark that $(\tilde{B}_i(t), t \geq 0)$ is also a Brownian motion. The matrix σ returns the correlated assets S_i.

5.2.2 *Infinitesimal generator and diffusion*

We shall see that with a diffusion process we can associate a linear, second-order partial differential operator. We call this operator the infinitesimal generator of the diffusion. We can show that a diffusion is a homogeneous Markov process and we shall see that this generator coincides with the notion of an infinitesimal generator of the semigroup associated with the homogeneous Markov process defined in Section 2.1.1.

For simplicity, we shall start by assuming that the diffusion evolves in **R**. We denote by $(X(t), t \geq 0)$ a solution of:

$$dX(t) = b\left(X(t)\right) \, dt + \sigma\left(X(t)\right) \, dB(t) \tag{5.8}$$

$(B(t), t \geq 0)$ being a Brownian motion with filtration $(\mathcal{F}_t, t \geq 0)$.

Proposition 5.2.3 *Let A be the differential operator which, to a function f of class C^2, associates the function:*

$$(Af)(x) = \frac{\sigma^2(x)}{2} f''(x) + b(x) f''(x).$$

Then, for every function f of class C^2 with bounded derivatives, the process $M(t) = f(X(t)) - \int_0^t Af(X(s)) \, ds$ is a martingale with respect to the filtration of the Brownian motion. In particular, we have, for all t:

$$\mathbf{E}\left(f(X(t))\right) = f(x) + \mathbf{E}\left(\int_0^t Af(X(s)) \, ds\right).$$

Proof The Itô formula gives:

$$f(X(t)) = f(X_0) + \int_0^t f'(X(s)) \, dX(s) + \frac{1}{2} \int_0^t f''(X(s)) \sigma^2(X(s)) \, ds.$$

From which:

$$f(X(t)) = f(X_0) + \int_0^t f'(X(s)) \sigma(X(s)) \, dB(s)$$
$$+ \int_0^t \left(\frac{1}{2} \sigma^2(X(s)) f''(X(s)) + b(X(s)) f'(X(s))\right) \, ds.$$

Since f has a bounded derivative, and $|\sigma(x)| < K(1 + |x|)$ we easily verify that, for all $t \geq 0$, $\mathbf{E}\left(\int_0^t |f'(X(s)) \sigma(X(s))|^2 \, ds\right) < +\infty$. Then, we use

Proposition 5.1.7 to deduce that the process $\left(\int_0^t \phi(s) \, dB(s), t \geq 0\right)$ is a martingale. This proves the first part of the proposition and we obtain the last result by taking the expected value. □

We can deduce Theorem 5.2.4 from this result.

Theorem 5.2.4 *We denote by $X^x(t)$ the solution of the stochastic differential equation (5.8) such that $X_0^x = x$. Then, if f is a function of class C^2 with bounded derivatives, the function $t \to \mathbf{E}(f(X^x(t)))$ is differentiable and we have:*

$$\frac{d}{dt}\mathbf{E}(f(X^x(t)))|_{t=0} = (Af)(x).$$

Remark The differential operator A is called *the infinitesimal generator* of the diffusion process $(X^x(t), t \geq 0)$. The theorem proves that it is a matter of the infinitesimal generator of the semigroup (in the sense of analysis) associated with X^x (see also, on this subject page 129) given by

$$P_t f(x) = \mathbf{E}(f(X_t^x)), \quad \text{for } t \geq 0, \quad x \in \mathbf{R}.$$

Proof If we denote by $X^x(t)$ the solution of the stochastic equation (5.8) such that $X_0^x = x$, we deduce from Proposition 5.2.3 that:

$$\mathbf{E}(f(X^x(t))) = f(x) + \mathbf{E}\left(\int_0^t Af(X^x(s)) \, ds\right).$$

Further, since the derivatives of f are bounded by a constant K_f and $|b(x)| + |\sigma(x)| \leq K(1 + |x|)$ we can show that:

$$\mathbf{E}\left(\sup_{s \leq T} |Af(X^x(s))|\right) < +\infty.$$

We can therefore apply the Lebesgue theorem ($x \mapsto Af(x)$ and $s \mapsto X^x(s)$ are continuous functions) to deduce that:

$$\frac{d}{dt}\mathbf{E}(f(X^x(t)))|_{t=0} = \lim_{t \to 0} \mathbf{E}\left(\frac{1}{t}\int_0^t Af(X^x(s)) \, ds\right) = Af(x).$$

□

This result generalizes to the case of vector valued diffusions with the stochastic differential equation:

$$\begin{cases} dX_1(t) = b_1(X(t)) \, dt + \sum_{j=1}^p \sigma_{1j}(X(t)) \, dB_j(t), \\ \vdots \quad \vdots \quad\quad\quad\quad\quad \vdots \\ dX_n(t) = b_n(X(t)) \, dt + \sum_{j=1}^p \sigma_{nj}(X(t)) \, dB_j(t). \end{cases} \quad (5.9)$$

Theorem 5.2.5 *We assume that the hypotheses of Theorem 5.2.2 are satisfied. We introduce the differential operator A which, to a function f of class $\mathcal{C}^2$ from $\mathbf{R}^n$ to $\mathbf{R}$, associates the function:*

$$(Af)(x) = \frac{1}{2} \sum_{i,j=1}^{n} a_{ij}(x) \frac{\partial^2 f}{\partial x_i \partial x_j}(x) + \sum_{j=1}^{n} b_j(x) \frac{\partial f}{\partial x_j}(x), \qquad (5.10)$$

where $a_{ij}(x) = \sum_{k=1}^{p} \sigma_{ik}(x)\sigma_{jk}(x)$ (with the matrix notation $a(x) = \sigma(x)\sigma^*(x)$ where $\sigma^*(x)$ is the transpose of the matrix $\sigma(x) = (\sigma_{ij}(x))_{i,j}$). We have:

$$\frac{d}{dt} \mathbf{E}\left(f\left(X^x(t)\right)\right)\big|_{t=0} = Af(x).$$

Remark We note that, for all x the matrix $a(x) = (a_{ij}(x))_{1 \le i,j \le n}$ is necessarily positive semidefinite but that nothing prevents it from being singular at particular points of $\mathbf{R}^n$. It is one of the strengths of the Monte-Carlo method that it adapts to these situations without great difficulty.
Proof With the help of the Itô formula (5.3), we obtain:

$$df(X(t)) = \sum_{i=1}^{n} \frac{\partial f}{\partial x_i}(X(t))\, dX_i(t) + \frac{1}{2} \sum_{i,j=1}^{n} \frac{\partial^2 f}{\partial x_i \partial x_j}(X(t))(\sigma\sigma^*)_{ij}(X(t))\, dt,$$

or

$$df(X(t)) = Af(X(t))\, dt + \sum_{i=1}^{n} \sum_{j=1}^{p} \frac{\partial f}{\partial x_i}(X(t))\sigma_{ij}(X(t))\, dB_j(t).$$

By integrating between 0 and t, and by taking account of the fact that, for all i and j:

$$\mathbf{E}\left(\int_0^t \frac{\partial f}{\partial x_i}(X(t))\sigma_{ij}(X(t))\, dB_j(t)\right) = 0,$$

we deduce that $\mathbf{E}(f(X(t))) = f(x) + \mathbf{E}\left(\int_0^t Af(X(s))\, ds\right)$. We finish by the same argument as in the proof of Theorem 5.2.4. □

We shall now show that the solutions of stochastic differential equations with infinitesimal generator A allow us to represent solutions of parabolic equations associated with the same operator.

5.2.3 *Diffusions and evolution problems*

To give a probabilistic interpretation to parabolic problems we use the Feynman–Kac formula. We shall establish this formula in a simple case.

In what follows, we restrict ourselves to showing that under regularity hypotheses on the solution of a partial differential equation this solution can be written in the form of the expected value of a functional of a diffusion. These results are not entirely satisfactory, in particular, we make very strong regularity hypotheses on the solutions *a priori*. These hypotheses can be partially relaxed (see Friedman, 1975; Bensoussan and Lions, 1978), but the explanation rapidly becomes very technical.

We assume that $b(x)$ and $\sigma(x)$ satisfy the hypotheses of Theorem 5.2.2, assuring existence and uniqueness of solutions of the stochastic differential equation (5.7).

Proposition 5.2.6 *If $u(t,x)$ is a function of class $C^{1,2}$ in (t,x) with derivative in x bounded, and $(X(t), t \geq 0)$ is a solution of (5.8), if $c(x)$ is a continuous function, bounded below over $\mathbf{R}^+ \times \mathbf{R}$, the process $(M(t), t \geq 0)$ with M_t equal to:*

$$e^{-\int_0^t c(X(s))\,ds} u(t, X(t)) - \int_0^t e^{-\int_0^s c(X(\eta))\,d\eta} \left(\frac{\partial u}{\partial t} + Au - cu \right) (s, X(s))\,ds$$

is a martingale with respect to the filtration $(\mathcal{F}_t, t \geq 0)$.

Proof With the help of the Itô formula (5.3), we obtain:

$$du(t, X(t)) = \left(\frac{\partial u}{\partial t}(t, X(t)) + Au(t, X(t)) \right) dt + \sum_{j=1}^d H_j(t)\,dB_j(t),$$

$H_j(t)$ being $\mathcal{F}_t$ measurable processes such that $\mathbf{E}\left(\int_0^T H_j(s)^2\,ds \right) < +\infty$ for all T.

On the other hand, we can differentiate the product $e^{-\int_0^t c(X(s))\,ds} u(t, X(t))$, by applying the Itô formula, to obtain:

$$d \left(e^{-\int_0^t c(X(s))\,ds} u(t, X(t)) \right)$$
$$= e^{-\int_0^t c(X(s))\,ds} \left(du(t, X(t)) - c(X(t)) u(t, X(t))\,dt \right).$$

From which:

$$d \left(e^{-\int_0^t c(X(s))\,ds} u(t, X(t)) \right) = e^{-\int_0^t c(X(s))\,ds} \left(\frac{\partial u}{\partial t} + Au - cu \right)(t, X(t))\,dt$$
$$+ \sum_{j=1}^d e^{-\int_0^t c(X(s))\,ds} H_j(t)\,dB_j(t).$$

Since c is bounded below, if we set:

$$K_j(t) = e^{-\int_0^t c(X(\eta))\,d\eta} H_j(t),$$

we have $\mathbf{E}\left(\int_0^T K_j(s)^2\,ds\right) < +\infty$. Therefore, the processes:

$$\int_0^t e^{-\int_0^s c(X(\eta))\,d\eta} H_j(s)\,dB_j(s),$$

are martingales, and we have the result. $\square$

The preceding result allows us to establish representation formulae for solutions of parabolic equations.

The Feynman–Kac formula.   We are given a homogeneous diffusion process (the coefficients b and σ do not depend explicitly on the time t) with values in $\mathbf{R}^n$, the solution of (5.7). We denote by $X^{t,x}$ the unique solution of

$$X(s) = x + \int_t^s b(X(u))\,du + \int_t^s \sigma(X(u))\,dB(u), \quad s \geq t$$

$X^{t,x}$ is the solution of the stochastic differential equation (5.7) starting from x at t. We shall write X^x for $X^{0,x}$. Let f be a function from $\mathbf{R}^n$ to $\mathbf{R}$ and c a bounded continuous function. The following result allows us to write the solution of a parabolic partial differential equation in the form of an expected value. It is this result that is called the Feynman–Kac formula.

Note that contrary to the case of Brownian motion (Proposition 5.1.11) we are treating here a parabolic equation with final condition. We shall later see how to restate this result for parabolic equations with initial condition.

Theorem 5.2.7 *Let f and g be continuous functions and c a function bounded below. Assume that u is a function of class $C^{1,2}$ in (t,x) with derivative in x bounded over $[0,T] \times \mathbf{R}^n$, satisfying:*

$$\begin{cases} \left(\dfrac{\partial u}{\partial t} + Au - cu\right)(t,x) = f(x) & \text{for} \ (t,x) \in [0,T] \times \mathbf{R}^n, \\ u(T,x) \hspace{2.4cm} = g(x) & \text{for} \ x \in \mathbf{R}^n. \end{cases}$$

If we denote by $\beta_{t,s} = e^{-\int_t^s c(X^{t,x}(\eta))\,d\eta}$, then for all $(t,x) \in [0,T] \times \mathbf{R}^n$:

$$u(t,x) = \mathbf{E}\left(\beta_{t,T}\,g(X^{t,x}(T)) - \int_t^T \beta_{t,s}\,f(X^{t,x}(s))\,ds\right).$$

Remark As the process is homogeneous, since b and σ do not depend explicitly on t, the law $(X^{t,x}(s), s \geq t)$ is identical to that of $(X^x(s), s \geq 0)$. We can also write the result of the preceding theorem in the form:

$$u(t,x) = \overline{\mathbf{E}}\left(\beta_{0,T-t}\,g(X^x(T-t)) - \int_0^{T-t} \beta_{0,s-t}\,f(X^x(s))\,ds\right).$$

Proof Let u be a solution of the preceding equation. By Proposition 5.2.6, we know that the process:

$$M(s) = \beta_{t,s}\, u(t, X^{t,x}(s)) + \int_t^s \beta_{t,v} \left(\frac{\partial u}{\partial t} + Au - cu \right)(v, X^{t,x}(v))\, dv$$

is a martingale for $s \geq t$. By writing $\mathbf{E}(M(t)) = \mathbf{E}(M(T))$, and as $\partial u / \partial t + Au - cu = f$ and $u(T, x) = g(x)$, we obtain:

$$u(t, x) = \mathbf{E}\left(\beta_{t,T}\, g(X^{t,x}(T)) - \int_t^T \beta_{t,s}\, f(X^{t,x}(s))\, ds \right).$$

$\square$

Remark When we set $c = 0$ and $f = 0$ in the Feynman–Kac formula, we obtain, if u is a regular function which is the solution of:

$$\begin{cases} u(T, x) = g(x), & \text{for} \quad x \in \mathbf{R}^n, \\ \left(\dfrac{\partial u}{\partial t} + Au \right)(t, x) = 0, & \text{for} \quad (t, x) \in [0, T] \times \mathbf{R}^n, \end{cases}$$

that:

$$u(t, x) = \mathbf{E}\left(g(X^{t,x}(T)) \right) = \mathbf{E}\left(g(X^x(T - t)) \right).$$

We therefore have a representation formula for a class of parabolic problems. This equation is traditionally called a retrograde Kolmogorov equation.

Remark When we are interested in evolution problems associated with an operator that is independent of time and a problem with an initial, rather than final, condition, we can re-express the theorem in the following way. Let u be a (regular) function satisfying:

$$\begin{cases} u(0, x) = g(x), & \text{for} \quad x \in \mathbf{R}^n, \\ \dfrac{\partial u}{\partial t} = Au - cu + f, & \text{in } [0, T] \times \mathbf{R}^n, \end{cases}$$

then for all $(t, x) \in [0, T] \times \mathbf{R}^n$:

$$u(t, x) = \mathbf{E}\left(\beta_{0,t} g(X^x(t)) - \int_0^t \beta_{0,s} f(X^x(s))\, ds \right),$$

it is enough to convince ourselves to fix t and to set, for $s \leq t$, $v(s, x) = u(t - s, x)$. It is then easy to establish a retrograde equation satisfied by

v where we know a probabilistic representation. We finish by using the homogeneity of the associated diffusion.

It is this last representation formula that allows us to recover the result of the same type (Proposition 5.1.11) as was obtained for Brownian motion.

5.2.4 *Stationary problems and diffusion*

We shall now give a probabilistic representation of the solutions of stationary equations of the type:

$$Au = f \quad \text{in} \quad D, \qquad u = g \quad \text{in} \quad \partial D,$$

where D is an open bounded set. We have the following theorem:

Theorem 5.2.8 *Let f and g be two bounded continuous functions defined over $\mathbf{R}^n$. Let D be a regular open set with boundary ∂D. If u is a continuous bounded function over $\bar{D}$ of class C^2 with bounded derivatives satisfying:*

$$\begin{cases} Au(x) = f(x) & \text{in } D, \\ u(x) \;\; = g(x) & \text{on } \partial D, \end{cases}$$

if $X^x(t)$ is the unique solution of:

$$X(t) = x + \int_0^t b(X(s))\,ds + \int_0^t \sigma(X(s))\,dB(s),$$

and if $\tau^x = \inf\{t > 0, X^x(t) \notin D\}$ is such that $\mathbf{E}(\tau^x) < +\infty$, then:

$$u(x) = \mathbf{E}\left(g(X^x(\tau^x))\right) - \mathbf{E}\left(\int_0^{\tau^x} f(X^x(s))\,ds\right).$$

Proof As u is of class C^2 with bounded derivatives:

$$M(t) = u(X^x(t)) - \int_0^t Au(X^x(s))\,ds$$

is a martingale. We further know that τ^x is a stopping time. By applying Proposition 5.1.7 at the bounded stopping time $\tau^x \wedge N$, we obtain:

$$u(x) = \mathbf{E}\left(u(X^x(\tau^x \wedge N))\right) - \mathbf{E}\left(\int_0^{\tau^x \wedge N} f(X^x(s))\,ds\right).$$

Now, as we have assumed that $\mathbf{E}(\tau^x) < +\infty$ and since f and u are bounded and continuous, we can pass to the limit as N tends to infinity. We then obtain:

$$u(x) = \mathbf{E}\left(g(X^x(\tau^x))\right) - \mathbf{E}\left(\int_0^{\tau^x} f(X^x(s))\,ds\right).$$

$\square$

Remark The hypotheses of the theorem are too strong. Their only virtue is to facilitate the proof. For more precise results we can consult Friedman (1975) or Bensoussan and Lions (1978).

Remark We see that for evolution problems or for stationary problems, it is not indispensable that the matrix $(a_{ij}(x))_{1\leq i,j\leq n}$ is positive definite for all x. This is one of the interesting factors for this type of method, in particular, for numerical calculations. Indeed in this case, classical finite element methods lead to the inversion of a linear system which is not strictly positive definite and which is therefore ill conditioned.

On the other hand, the probabilistic representation formula allows us sometimes to show results about partial differential equations, in particular, in degenerate cases. We can, for example, prove the convergence of (nonprobabilistic) numerical schemes with the help of probabilistic techniques. We can find some examples of these techniques in Kushner (1977, 1990).

5.2.5 *Diffusion and the Fokker–Planck equation*

As in the case of transport processes, we can write an equation satisfied by the density of the the diffusion law at the time t. This equation is known as the Fokker–Planck equation. Let $(X(t), t \geq 0)$ be the unique solution of:

$$dX(t) = b(X(t))\,dt + \sigma(X(t))\,dB(t), \quad X(0) = X,$$

b and σ being two Lipschitz functions and X a square integrable random variable that is $\mathcal{F}_0$-measurable having density $p_0(x)$. The infinitesimal generator A of this diffusion is given by

$$Af(x) = \frac{1}{2}\sum_{i,j=1}^n a_{ij}(x)\frac{\partial^2 f}{\partial x_i \partial x_j}(x) + \sum_{j=1}^n b_j(x)\frac{\partial f}{\partial x_j}(x),$$

where $a_{ij}(x) = \sum_{k=1}^p \sigma_{ik}(x)\sigma_{jk}(x)$.

As in the case of transport processes (see Section 2.3.1) the following equation, which is easy to establish, is the basis of the Fokker–Planck equation. In what follows, we denote by μ_t the law $X(t)$:

$$\mu_t(f) = \mathbf{E}\left(f(X(t))\right), \quad \text{for } f \text{ continuous and bounded.}$$

Theorem 5.2.9 *Let f be a bounded function of class C^2 having first- and second-order bounded partial derivatives. Then:*

$$\mu_t(f) = \mu_0(f) + \int_0^t \mu_s(Af)\, ds.$$

Proof Since f is of class C^2 and has bounded derivatives, an application of the Itô formula (see Proposition 5.2.3 in the case $n = p = 1$) shows that:

$$M(t) = f(X(t)) - f(X(0)) - \int_0^t Af(X(s))\, ds,$$

is a martingale. By writing that $\mathbf{E}\,(M(t)) = 0$, we obtain:

$$\mathbf{E}\,(f(X(t))) = \mathbf{E}\,(f(X(0))) + \mathbf{E}\left(\int_0^t Af(X(s))\, ds\right).$$

The Fubini theorem confirms that:

$$\mathbf{E}\left(\int_0^t Af(X(s))\, ds\right) = \int_0^t \mathbf{E}\,(Af(X(s))\, ds),$$

and we have completed the proof. □

We now define the adjoint operator A^* of A by

$$A^* f(x) = \frac{1}{2} \sum_{i,j=1}^{n} \frac{\partial^2 (a_{i,j}(x)f(x))}{\partial x_i \partial x_j} - \sum_{j=1}^{n} \frac{\partial (b_j(x)f(x))}{\partial x_j}.$$

Note that if f and g are two functions of class C^2 where at least one has compact support, we have:

$$\int_{\mathbf{R}^n} (Af)(x)g(x)\, dx = \int_{\mathbf{R}^n} f(x)(A^*g)(x)\, dx.$$

We then have the following result:

Corollary 5.2.10 *Assume that the law of the random variable $X(t)$ has a density $p(t,x)$ of class $C^{1,2}$, then this density satisfies the equation:*

$$\begin{cases} \dfrac{\partial p}{\partial t}(t,x) = (A^*p)(t,x) & \text{for } t \geq 0,\ x \in \mathbf{R}^n, \\ p(0,x) = p_0(x) & \text{almost surely in } x. \end{cases}$$

Proof Let f be a regular function with compact support, we have, from Theorem 5.2.9:

$$\int_{\mathbf{R}^n} f(x)p(t,x)\,dx = \int_{\mathbf{R}^n} f(x)p_0(x)\,dx + \int_0^t \int_{\mathbf{R}^n} Af(x)p(s,x)\,dx.$$

As $f(.)$ and $p(t,.)$ are regular, and f has compact support, we deduce that:

$$\int_{\mathbf{R}^n} Af(x)p(s,x)\,dx = \int_{\mathbf{R}^n} f(x)A^*p(s,x)\,dx.$$

Therefore,

$$\int_{\mathbf{R}^n} f(x)p(t,x)\,dx = \int_{\mathbf{R}^n} f(x)p_0(x)\,dx + \int_0^t ds \int_{\mathbf{R}^n} f(x)A^*p(s,x)\,dx.$$

We deduce the result by using the regularity of p. □

Remark Note that, without any regularity hypothesis, p is a solution in a weak sense of the Fokker–Planck equation, as we have, for every function $f \in C^\infty$ with compact support:

$$\int_{\mathbf{R}^n} f(x)p(t,x)\,dx = \int_{\mathbf{R}^n} f(x)p_0(x)\,dx + \int_0^t ds \int_{\mathbf{R}^n} (Af)(x)p(s,x)\,dx.$$

We can further define a distribution A^*p, for every measurable positive function p, by duality, by setting, if f is a function C^∞ with compact support:

$$\langle A^*p, f \rangle = \int_{\mathbf{R}^n} p(x)(Af)(x)\,dx.$$

With this definition, we see that we have:

$$\int_{\mathbf{R}^n} f(x)p(t,x)\,dx = \int_{\mathbf{R}^n} f(x)p_0(x)\,dx + \int_0^t ds \langle A^*p(s,.), f \rangle.$$

This a way of writing, in the sense of distributions:

$$p(t,.) = p_0(.) + \int_0^t A^*p(s,.)\,ds.$$

Stationary Fokker–Planck equation. Note that when the process has an invariant probability law $\mu_0(dx)$ (that is if, for all $t \geq 0$, and for every measurable function f, $\mu_t(f) = \mathbf{E}\left(f(X(t))\right) = \mu_0(f))$, an application of Theorem 5.2.9 shows that, for f a regular function, we have:

$$\mu_0(Af) = 0.$$

If we suppose further that this invariant law has a regular density $p_0(x)$ we see that it must satisfy the equation:

$$(A^*p_0)(x) = 0.$$

This equation is called the *Stationary Fokker–Planck equation*.

Example 9 In one dimension the operator A^* takes the form:

$$A^*f = \frac{1}{2}\frac{\partial^2(\sigma^2(x)f(x))}{\partial x^2} - \frac{\partial(b(x)f(x))}{\partial x}.$$

Therefore, for the Ornstein–Uhlenbeck process, the solution of the equation:

$$dX(t) = -cX(t)\,dt + \sigma\,dB(t),$$

the operator A^* becomes:

$$A^* = \frac{1}{2}\sigma^2\frac{\partial^2 f(x)}{\partial x^2} + c\frac{\partial(xf(x))}{\partial x}.$$

The Fokker–Planck equation is therefore written as

$$\begin{cases} \dfrac{\partial p}{\partial t}(t,x) = \dfrac{1}{2}\sigma^2\dfrac{\partial^2 p(t,x)}{\partial x^2} + c\dfrac{\partial(xp(t,x))}{\partial x}, \\ p(0,x) = p_0(x). \end{cases}$$

The stationary Fokker–Planck equation takes the form:

$$\frac{1}{2}\sigma^2\frac{\partial^2 p_0(x)}{\partial x^2} + c\frac{\partial(xp_0(x))}{\partial x} = 0.$$

Remark It is easy to verify that the function $f(x) = e^{-cx^2/\sigma^2}$ is a solution of the stationary Fokker–Planck equation. This strongly suggests that the centred Gaussian law with variance $\sigma^2/2c$ is *invariant* for this process: if $X(0)$ follows the preceding law, the marginal law $X(t)$ remains equal to that of $X(0)$. This is effectively the case and this can be proved, for example, by direct reasoning on the process.

Example 10 Let U be a function from $\mathbf{R}^n$ to $\mathbf{R}^+$. We shall denote by $\nabla U(x)$ its gradient at the point x. When we consider the solution of the equation:

$$dX(t) = -\nabla U(X(t))\,dt + \sigma\,dB(t),$$

the operator A^* is written as

$$A^* f(x) = \frac{1}{2}\sigma^2 \frac{\partial^2 f(x)}{\partial x^2} + \frac{\partial(\nabla U(x)f(x))}{\partial x}.$$

It is then easy to show that if U is regular, the function:

$$f_0(x) = C e^{-2U(x)/\sigma^2},$$

satisfies $A^* f_0(x) = 0$. If U tends quickly enough to $+\infty$ as $|x|$ tends to $+\infty$, we can find a constant C such that $\int_{\mathbf{R}^n} f_0(x)\,dx = 1$. In this case, the probability $f_0(x)\,dx$ is also an invariant law for this process. We recover Example 10 by setting $U(x) = c|x|^2/2$.

5.2.6 Applications in financial mathematics

The Feynman–Kac formula has been very useful in the domain of financial mathematics since the 1970s. We give several examples of its use in this section.

5.2.6.1 An example of option pricing.
We start with a simple example. Imagine that we want to calculate the price of a European option that promises $f(S(T))$ at time T in the Black–Scholes model. This means that $(S(t), t \geq 0)$ is the unique solution of:

$$dS(t) = S(t)\left(r\,dt + \sigma\,dB(t)\right), \quad S(0) = x$$

r and σ being real positive numbers, $(B(t), t \geq 0)$ a Brownian motion, and f being a given function. When $f(x) = (x - K)_+$ we speak of a call and when $f(x) = (K - x)_+$ of a put. One of these problems (see, e.g. Lamberton and Lapeyre, 1991, for more precise information) is to calculate the price of this option expressed in the form:

$$\mathbf{E}\left(e^{-rT} f(S(T))\right).$$

By using Theorem 5.2.7, we see that this price is equal to $u(0, x)$ if u is a regular solution of

$$\begin{cases} \dfrac{\partial u}{\partial t} + \dfrac{\sigma^2}{2}x^2\dfrac{\partial^2 u}{\partial x^2} + rx\dfrac{\partial u}{\partial x} - ru = 0 & \text{in } [0,T] \times]0, +\infty[\\ u(T, x) = f(x), \quad \forall x \in]0, +\infty[. \end{cases}$$

The price of this option is therefore expressed with the help of a solution of a partial differential equation.

Note that we have seen in Example 7 that, by setting:

$$g(t,y) = x \exp\left(\left(r - \frac{\sigma^2}{2}\right)t + \sigma y\right),$$

$S(t)$ can be written $S(t) = g(t, B(t))$. The price of the option therefore takes the form $\mathbf{E}\left(e^{-rT}g(T, B(T))\right)$. This expression lends itself to a Monte-Carlo type calculation (and moreover to many other numerical methods!). We have already recalled this type of problem in Chapter 1.

Remark Note that the case cited is very suitable for a simulation method. In effect, we must simulate a function of the Brownian motion at the final time $e^{-rT}g(T, B(T))$. Since the simulation of $B(T)$ for a unique fixed T can be made easily in an accurate and efficient way, this does not pose any problem. In many cases, on the one hand, we do not know how to simulate the process exactly (as in the case where we have a complicated diffusion process); on the other hand, we have to approximate a functional $\psi(X(s), s \geq 0)$ by a quantity $\psi_n(X(t_1), \ldots, X(t_n))$, which only depends on a finite number of values of X. This is not always easy. This is the case, for example, when we want to evaluate the average price of an option, that is, when we want to evaluate $\mathbf{E}\left(e^{-rT}f(I(T))\right)$, with $I(T) = (1/T)\int_0^T S(s)\,ds$ where $(S(s), s \geq 0)$ follows the Black–Scholes model.

An option calculation in many dimensions. We assume that the asset model is that described in Example 8. In this framework, we can show that an option promise $f(S_1(T), \ldots, S_N(T))$ at time T becomes:

$$P = \mathbf{E}\left(e^{-rt}f(S_1(T), \ldots, S_N(T))\right).$$

A classical example of this type of situation is the price of an indexed option. The index is then a balance of asset prices $S_i(t)$:

$$I(t) = a_1 S_1(t) + \cdots + a_N S_N(t)$$

$(a_i, 1 \leq i \leq N)$ being positive real numbers with sum 1. A call on the index $I(t)$, that is, an option promise $(I(T) - K)_+$ at time T corresponds to the following choice of f:

$$f(x_1, \ldots, x_N) = (a_1 x_1 + \cdots + a_N x_N - K)_+.$$

The problem of calculating P is linked to a partial differential equation (in $\mathbf{R}^N$). By using the Feynman–Kac Theorem 5.2.7, we can show that if

$$Af(x) = \sum_{i=1}^{N} r x_i \frac{\partial f}{\partial x_i}(x) + \frac{1}{2}\sum_{i,j=1}^{N} a_{ij}x_i x_j \frac{\partial^2 f}{\partial x_i x_j}(x),$$

where $a_{ij} = \sum_{k=1}^{N} \sigma_{ik}\sigma_{jk}$, and if u is regular solution of the partial differential equation:

$$\begin{cases} \dfrac{\partial u}{\partial t} + Au - ru = 0 \quad \text{in } [0, T] \times]0, +\infty[^N \\ u(T, x) = f(x), \quad \forall x \in]0, +\infty[^N, \end{cases}$$

then $P = u(0, S(0))$.

Remark We note that in this case we do not know a numerical method to evaluate exactly the price of this option: there is no explicit formula, and numerical methods are not suitable because of the dimension (the asset number N is often greater than 40 and can be several hundred!) of the parabolic problem to be resolved. Resorting to a Monte-Carlo method (or to a simplification of the model) is indispensable in this case.

Calculation of the average rate of an option. On average an option promises, at time T, the amount:

$$\left(\frac{1}{T} \int_0^T S(s)\, ds - K \right)_+.$$

Its price M can be expressed in the form:

$$M = \mathbf{E} \left(e^{-rT} \left(\frac{1}{T} \int_0^T S(s)\, ds - K \right)_+ \right),$$

with $(S(t), t \geq 0)$ which follows the Black–Scholes model:

$$dS(t) = S(t)\, (r\, dt + \sigma\, dB(t)), \quad S(0) = x,$$

where $(B(t), t \geq 0)$ is a Brownian motion with probability $\mathbf{P}$. We know that we can then write $S(t)$ in the form:

$$S(t) = x \exp \left(\left(r - \frac{\sigma^2}{2} \right) t + \sigma B(t) \right).$$

We shall denote $I(t) = \int_0^t S(s)\, ds$. Note that the pair $(S(t), I(t))$ is a diffusion, in effect:

$$\begin{cases} dS(t) = rS(t)\, dt + \sigma S(t)\, dB(t), \\ dI(t) = S(t)\, dt. \end{cases}$$

The infinitesimal generator associated with this diffusion is given by

$$Af(x, y) = \frac{\sigma^2}{2} x^2 \frac{\partial^2 f}{\partial x^2} + rx \frac{\partial f}{\partial x} + x \frac{\partial f}{\partial y}.$$

By using Theorem 5.2.7, we can confirm that if $u(t, x, y)$ is a regular solution of the partial differential equation:

$$\begin{cases} \dfrac{\partial u}{\partial t} + Au - ru = 0 & \text{in } [0, T] \times]0, +\infty[^2 \\ u(T, x, y) = f(x, y), & \forall x \in]0, +\infty[^2, \end{cases}$$

with $f(x, y) = ((y/T) - K)_+$, then:

$$M = u(0, x, 0) = \mathbf{E}\left(e^{-rT} f\left(S(T), \int_0^T S(s)\, ds \right) \right).$$

We therefore see that the calculation of the price of an average option rate leads to the solution of a parabolic equation in $\mathbf{R}^2$. This equation is however strongly degenerate, which makes its numerical solution difficult.

We can also represent the price of an average option rate with the help of a parabolic equation in one dimension (this representation is taken from Rogers and Shi, 1995).

Proposition 5.2.11 *If $u(t, x)$ is a regular solution of the partial differential equation:*

$$\begin{cases} \dfrac{\partial u}{\partial t} - \left(\dfrac{1}{T} + rx \right) \dfrac{\partial u}{\partial x} + \dfrac{1}{2}\sigma^2 x^2 \dfrac{\partial^2 u}{\partial x^2} = 0 & \text{in } [0, T] \times \mathbf{R}, \\ u(T, x) = \max(0, -x), & \forall x \in \mathbf{R}, \end{cases}$$

then the price of the average option rate becomes $M = xu(0, K/x)$.

Proof For this we remark that, if we set:

$$\xi(t) = \frac{K - (1/T) \int_0^t S(s)\, ds}{S(t)}$$

$(\xi(t), t \geq 0)$ is (by applying the Itô formula) the solution of

$$d\xi(t) = -\frac{1}{T}\, dt + \xi(t)(-\sigma\, dB(t) - r\, dt + \sigma^2\, dt).$$

As $S_T = \exp\left(rT + \sigma B(T) - (\sigma^2 T)/2 \right)$ we can express the price M in the form (denoting $x_- = \max(0, -x)$):

$$M = \mathbf{E}\left(e^{-rT} S(T)\, (\xi(T))_- \right) = \mathbf{E}\left(e^{\sigma B(T) - (1/2)\sigma^2 T}\, (\xi(T))_- \right).$$

Or, by using the Girsanov Theorem 5.4.3, we see that under the probability $\tilde{P}$ defined for measurable random variables $\mathcal{F}_T$, by

$$d\tilde{\mathbf{P}} = e^{\sigma B(T) - (1/2)\sigma^2 T}\, d\mathbf{P},$$

$\tilde{B}(t) = B(t) - \sigma t$ is a Brownian motion. We therefore have:

$$d\xi(t) = -\left(\frac{1}{T} + r\xi(t)\right) dt - \xi(t)\sigma \, d\tilde{B}(t),$$

and

$$M = \tilde{\mathbf{E}}\left((\xi(T))_-\right).$$

The generator of the diffusion $\xi(t)$ under $\tilde{\mathbf{P}}$ being:

$$-\left(\frac{1}{T} + rx\right)\frac{\partial u}{\partial x} + \frac{1}{2}\sigma^2 x^2 \frac{\partial^2 u}{\partial x^2},$$

an application of the Feynman–Kac formula leads to the result.   □

5.3   Simulation of diffusion processes

In this part, we shall show how we can implement Monte-Carlo methods, starting from simple cases (Brownian motion) then generalizing to diffusions.

We shall see that the solutions of some evolutionary or elliptic partial differential equations are represented in the form of the expected value of a functional of a diffusion. We are then led to calculating quantities of the type:

$$\mathbf{E}\left(\psi(X(s), s \ge 0)\right),$$

where $(X(s), s \ge 0)$ is the solution of a stochastic differential equation. To implement a Monte-Carlo method starting from the representation above, we then:

- simulate the trajectory of the process. Note that we can only computationally simulate this trajectory at a *finite* number of times $0 \le t_1 < t_2 < \cdots < t_n$. We do this, when we do not know that solution explicitly, with the help of approximations to stochastic differential equations.

- approximate $\psi(X(s), s \ge 0)$ by a quantity of the type ψ_n, a unique function of $X(t_1), \ldots, X(t_n)$. This stage is all the more delicate as the functional ψ is complicated (for example if it involves stopping times, ...).

The case of the Brownian motion.   The simplest diffusion is real Brownian motion. It is the basis of the simulation of more complex diffusions. The simulation of the Brownian motion $(B(t), 0 \le t \le T)$ at a given time T is very easy since the law of $B(T)$ is Gaussian centred reduced with variance T. Further, as Brownian motion has independent increments, we can reduce the simulation of the path at the times $t_1 < t_2 < \cdots < t_n$ to the preceding problem by setting:

$$B(t_1) = \sqrt{t_1}\, g_1,$$
$$\cdots$$
$$B(t_k) - B(t_{k-1}) = \sqrt{t_k - t_{k-1}}\, g_k,$$
$$\cdots$$
$$B(t_n) - B(t_{n-1}) = \sqrt{t_n - t_{n-1}}\, g_n,$$

where $(g_1, \ldots, g_n)$ are n independent centred Gaussian with variance 1. Note that, proceeding in this way, we can exactly simulate the law of the n-tuplet $(B(t_1), B(t_2), \ldots, B(t_n))$. This will no longer be the case for the simulation of stochastic differential equations discussed later in this chapter.

When we want to simulate a solution of a stochastic differential equation and we do not know how to express this solution in a simple way with the help of Brownian motion, we have to resort to approximations. We shall state, now, the simplest methods of simulation of solutions of stochastic differential equations: the Euler scheme and the Milshtein scheme.

We assume, in what follows, that we want to calculate the solution of a parabolic problem represented with the help of the diffusion process with values in $\mathbf{R}^n$ the solution of:

$$dX(t) = b(X(t))\, dt + \sigma(X(t))\, dB(t), \quad X(0) = x,$$

b being a function from $\mathbf{R}^n$ to $\mathbf{R}^n$, σ a mapping from $\mathbf{R}^n$ to $\mathbf{R}^{n \times p}$, and $(B(t), t \geq 0)$ a p-dimensional Brownian motion.

5.3.1   The Euler scheme

The simplest scheme is the Euler scheme. We are given a strictly positive time step h and we calculate an approximation $\bar{X}$ of the process X at the time $t_k = kh$, $k \geq 0$, by setting $\bar{X}(0) = x$ and for $k > 1$:

$$\bar{X}((k+1)h) = \bar{X}(kh) + b\left(\bar{X}(kh)\right) h + \sigma\left(\bar{X}(kh)\right) (B((k+1)h) - B(kh)).$$
$$(5.11)$$

This is the natural generalization, to stochastic differential equations, of the Euler scheme that we use for ordinary differential equations. Note that the random variables $(B((k+1)h) - B(kh), k \geq 0)$ are centred Gaussian random variables with variance–covariance matrix $h\mathrm{Id}$ (Id being the identity matrix of $\mathbf{R}^p$). This sequence of random variables is very easy to simulate, the computational implementation of this scheme is simple.

Theorem 5.3.1 states a convergence result for this scheme and evaluates the precision of the approximation.

Theorem 5.3.1 *Let b and σ be two Lipschitz functions. Let $(B(t), t \geq 0)$ be a p-dimensional Brownian motion. We denote by $(X(t), t \geq 0)$ the* *unique solution of:*

$$dX(t) = b(X(t)) \, dt + \sigma(X(t)) \, dB(t), \quad X(0) = x$$

and $(\bar{X}(kh), k \geq 0)$ the sequence of random variables defined by Equation (5.11). Then, for all $q \geq 1$:

- $\mathbf{E}\left(\sup_{k, kh \leq T} \left| X(kh) - \bar{X}(kh) \right|^{2q}\right) \leq Ch^q;$

- *in particular, if $h = T/N$, $\mathbf{E}\left(\left| X(T) - \bar{X}(T) \right|^{2q}\right) \leq Ch^q;$*

- *further, for all $\alpha < 1/2$, almost surely:*

$$\lim_{h \to 0} h^\alpha \sup_{k, kh \leq T} \left| \bar{X}(kh) - X(kh) \right| = 0;$$

- *finally, if b and σ are functions of class C^4 having bounded derivatives up to order 4, if f is a function of class C^4 having derivatives up to order 4 with polynomial growth, then, if $h = T/N$, there exists a constant C_T such that:*

$$\left| \mathbf{E}\left(f(X(T))\right) - \mathbf{E}\left(f(\bar{X}(T))\right) \right| \leq \frac{C_T}{N}.$$

Remark Theorem 5.3.1 shows that the speed of convergence in L^2 is of order $h^{1/2}$ and that the speed of almost surely convergence is of order $h^{1/2 - \epsilon}$, for all $\epsilon > 0$. The second result confirms that for very regular functions, the speed of convergence in law of the scheme is of order h.

We can find a detailed proof of this result in Talay (1986) and Faure (1992a,b).

5.3.2   The Milshtein scheme

There exist some notable improvements of the Euler scheme for ordinary differential equations (Runge Kutta, etc.). Many higher-order schemes have been proposed for stochastic differential equations, but they are often difficult to implement. For this we can consult Pardoux and Talay (1985), Kloeden and Platen (1992), and Talay (1995).

The best known of these schemes is the Milshtein scheme. It allows us (very often) to obtain a convergence path of order $h^{1-\epsilon}$, $\epsilon > 0$.

The case of one dimension. We start by explaining this in the case where $n = p = 1$. We proceed in the following way, we set $\bar{X}(0) = x$ and for $k > 1$:

$$\bar{X}((k+1)h) = \bar{X}(kh) + b\left(\bar{X}(kh)\right)h + \sigma\left(\bar{X}(kh)\right)\left(B((k+1)h) - B(kh)\right)$$

$$+ \sigma'(\bar{X}(kh))\sigma(\bar{X}(kh)) \int_{kh}^{(k+1)h} \left(B(s) - B(kh)\right) dB(s).$$

$$(5.12)$$

Remark We can understand the appearance of the supplementary term (relative to an Euler scheme) by considering what happens for the equation:

$$dX(t) = \sigma(X(t))\, dB(t).$$

The Euler schemes can be extended, by noting $t_k = kh$, at every time t from $[t_k, t_{k+1}]$ by setting:

$$\bar{X}(t) = \bar{X}(t_k) + \sigma\left(\bar{X}(t_k)\right)(B(t) - B(t_k)).$$

$\bar{X}(t)$ provides an approximation of $X(t)$, on the interval $[t_k, t_{k+1}]$, of $X(t)$ which is better than $\bar{X}(t_k)$, we can therefore hope that $\sigma(\bar{X}(t))$ will be a better approximation of $\sigma(X(t))$ than $\sigma(\bar{X}(t_k))$. A natural higher-order scheme is therefore:

$$\bar{X}(t) = \bar{X}(t_k) + \int_{t_k}^{t} \sigma\left(\bar{X}(s)\right)\, dB(s).$$

But, to first approximation:

$$\sigma\left(\bar{X}(t)\right) = \sigma\left(\bar{X}(t_k) + \sigma\left(\bar{X}(t_k)\right)(B(t) - B(t_k))\right)$$
$$\approx \sigma\left(\bar{X}(t_k)\right) + \sigma'\left(\bar{X}(t_k)\right)\sigma\left(\bar{X}(t_k)\right)(B(t) - B(t_k)).$$

This leads to the scheme:

$$\bar{X}(t) = \bar{X}(t_k) + \sigma\left(\bar{X}(t_k)\right)(B(t) - B(t_k))$$
$$+ \sigma\left(\bar{X}(t_k)\right)\sigma'\left(\bar{X}(t_k)\right)\int_{t_k}^{t}(B(s) - B(t_k))\, dB(s).$$

We therefore recover the Milshtein scheme when the derivative b is zero. As the dominant term in the error is linked to the Brownian motion it is not surprising that the derivative term b does not add a correction term of the same order of magnitude.

From the practical point of view it is crucial to note that we can obtain, by using the Itô formula:

$$\int_{kh}^{(k+1)h}(B(s) - B(kh))\, dB(s) = \frac{1}{2}\left((B((k+1)h) - B(kh))^2 - h\right).$$

This allows us to rewrite the Milshtein scheme in the form:

$$\bar{X}((k+1)h) = \bar{X}(kh) + \left(b\left(\bar{X}(kh)\right) - \tfrac{1}{2}\sigma'(\bar{X}(kh))\sigma(\bar{X}(kh))\right)h$$
$$+ \sigma\left(\bar{X}(kh)\right)(B((k+1)h) - B(kh))$$
$$+ \tfrac{1}{2}\sigma'(\bar{X}(kh))\sigma(\bar{X}(kh))(B((k+1)h) - B(kh))^2.$$

We see, in this form, that the scheme is easy to implement, since it is enough to know how to simulate the sequence $(B((k+1)h) - B(kh), k \geq 0)$.

Example 11 We give a simple example of the use of the Euler and Milshtein schemes. We consider the Black–Scholes process:

$$dS_t = S_t \left(r \, dt + \sigma \, dB(t) \right), \quad S(0) = x.$$

If we set $\Delta B_k = B((k+1)h) - B(kh)$, the Euler scheme takes the form:

$$\bar{X}((k+1)h) = \bar{X}(kh) \left(1 + rh + \sigma \Delta B_k \right).$$

The Milshtein scheme is written as

$$\bar{X}((k+1)h) = \bar{X}(kh) \left(1 + \left(r - \tfrac{1}{2}\sigma^2 \right) h + \sigma \Delta B_k + \tfrac{1}{2}\sigma^2 (\Delta B_k)^2 \right).$$

The case of dimension greater than one. When p, the dimension of the Brownian motion leading to the stochastic differential equation, is greater than one, this method is more delicate to implement. The scheme is presented, then, in the following way:

$$\bar{X}((k+1)h) = \bar{X}(kh) + b\left(\bar{X}(kh) \right) h + \sigma \left(\bar{X}(kh) \right) \left(B((k+1)h) - B(kh) \right)$$
$$+ \sum_{j,l=1}^{p} \left(\partial \sigma_j \sigma_l \right) \left(\bar{X}(kh) \right) \int_{kh}^{(k+1)h} \left(B_j(s) - B_j(kh) \right) dB_l(s),$$

$$(5.13)$$

with, for $1 \le i \le n$:

$$\left(\partial \sigma_j \sigma_l \right)_i = \sum_{r=1}^{n} \frac{\partial \sigma_{ij}}{\partial x_r} \sigma_{rl}.$$

Remark This form of the Milshtein scheme is difficult to simulate. In effect, it is necessary to be able to simulate the vector form of the random variables:

$$\left(B_j((k+1)h) - B_j(kh), \int_{kh}^{(k+1)h} \left(B_j(s) - B_j(kh) \right) dB_l(s) \right)$$

for $1 \le j \le p, 1 \le l \le p$. When $p = 2$, this reduces to knowing how to simulate the triplet:

$$\left(B_1(h), B_2(h), \int_0^h \left(B_1(s) \, dB_2(s) - B_2(s) \, dB_1(s) \right) \right).$$

Now we do not know, currently, how to simulate this triplet efficiently.

The Milshtein scheme is used essentially when it is possible to avoid eliminating the preceding problem. This recovers the one-dimensional case, but it is also used when the following commutativity condition is satisfied:

(C)   For all j, k in $\{1, \ldots, p\}$ and for all $x \in \mathbf{R}^n$:
$$\partial \sigma_j(x)\sigma_k(x) = \partial \sigma_k(x)\sigma_j(x).$$

Under hypothesis (C), it is easy to verify that we can rewrite the Milshtein scheme in the form:

$$\bar{X}((k+1)h) = \bar{X}(kh) + \left(b\left(\bar{X}(kh)\right) - \frac{1}{2} \sum_{j=1}^{p} (\partial\sigma_j\sigma_j)(\bar{X}(kh)) \right) h$$
$$+ \sigma\left(\bar{X}(kh)\right) \left(B((k+1)h) - B(kh)\right)$$
$$+ \frac{1}{2} \sum_{j,l=1}^{p} (\partial\sigma_j\sigma_l)\left(\bar{X}(kh)\right)$$
$$\times \left(B_j((k+1)h) - B_j(kh)\right)\left(B_l((k+1)h) - B_l(kh)\right).$$

We see that the simulation of this scheme then reduces to that of random variables:
$$((B_j((k+1)h) - B_j(kh)), k \geq 0, 1 \leq j \leq p).$$

Theorem 5.3.2 makes clear the rate of convergence of this scheme.

Theorem 5.3.2 *We assume that b and σ are functions that are twice continuously differentiable with bounded derivatives. We denote by $(X(t), t \geq 0)$ the unique solution of the equation:*

$$dX(t) = b(X(t))\, dt + \sigma(X(t))\, dB(t), \quad X(0) = x$$

and $(\bar{X}(kh), k \geq 0)$ the sequence of random variables defined by the Milshtein scheme (5.13). Then:

- *for all $q \geq 1$ $\sup_{k,kh \leq T} \mathbf{E}\left(\left|X(kh) - \bar{X}(kh)\right|^q\right) \leq Ch^q$;*
- *in particular, if $h = T/N$, $\mathbf{E}\left(\left|X(T) - \bar{X}(T)\right|^q\right) \leq Ch^q$;*
- *for all $\alpha < 1$, we have:*

$$\lim_{h \to 0} \frac{1}{h^\alpha} \sup_{k,kh \leq T} \left|X(kh) - \bar{X}(kh)\right| = 0.$$

Further, if b and σ are functions of class C^4 having bounded derivatives of order up to 4 and if f is a function of class C^4 having derivatives up to order 4, which are increasing polynomially, then, if $h = T/N$, there exists a constant $C_T > 0$ such that:

$$\left|\mathbf{E}\left(f(X(T))\right) - \mathbf{E}\left(f(\bar{X}(T))\right)\right| \leq \frac{C_T}{N}.$$

Remark Note that the result is true without the commutativity hypothesis (C).

For a proof of this result see Talay (1986) and Faure (1992*a,b*).

The Milshtein scheme improves the speed of *trajectorial* convergence of the Euler scheme since this speed is of order h in L^2 and of order $h^{1-\epsilon}$ almost surely, for all $\epsilon > 0$. Conversely, the speed of convergence in *law* for regular functions remains the same, that is, of order h. When we use Monte-Carlo methods the important speed of convergence is that of the convergence in law. From this point of view, the preceding result seems to indicate that the Milshtein scheme does not significantly improve the order of convergence of the Euler scheme for functionals of the type $f(X(T))$.

Other methods. We have just seen that it is easy to obtain a speed of convergence of order $\sqrt{h}$ using the Euler scheme and of order h thanks to the Milshtein scheme. We are trying to construct more precise schemes of higher order. However, a result due to Clark and Cameron (see Clark and Cameron (1980) or Faure (1992*b*)) proves, in the sense of the norm L^2, the quasi-optimality of the Euler scheme among all the schemes using only random variables $(B(kh), k \geq 0)$. Likewise, we can show that, in one dimension or under hypothesis (C), under the same conditions, the Milshtein scheme is quasi-optimal.

We can, however, construct generalizations of the Milshtein scheme having arbitrary trajectorial order of convergence, but these schemes use iterated integrals of the Brownian motion of order greater than we can simulate without great difficulty (see Kloeden and Platen, 1992).

We can consult Kloeden and Platen (1992) and Talay (1995), for a complete description of these other methods of discretization of stochastic differential equations. We shall find the proofs of the theorems that we have stated in Talay (1986), Faure (1992*a,b*), or Kloeden and Platen (1992).

5.4   Methods for variance reduction

Some variance reduction techniques presented in Chapter 1 can be exploited efficiently for diffusion processes. We shall study control variable and importance function techniques here. For a deeper treatment of these questions we can consult Newton (1994) and Wagner (1988). This section mostly follows Newton (1994).

5.4.1   *Control variables and predicted representation*

We want to evaluate, by a Monte-Carlo method, an expected value of the type:

$$\mathbf{E}(Z),$$

where $Z = \psi(X(s), 0 \leq s \leq T)$ and $(X(s), s \geq 0)$, which is solution of the stochastic differential equation:

$$dX(t) = b(X(t))\,dt + \sigma(X(t))\,dB(t), \quad X(0) = x.$$

To simplify the notation, we assume that the processes $(X(t), t \geq 0)$ and $(B(t), t \geq 0)$ take their values in $\mathbf{R}$. We recall that the control variable technique (see Section 1.4) consists of subtracting a random variable Y with explicitly calculable expected value:

$$\mathbf{E}(Z) = \mathbf{E}(Z - Y) + \mathbf{E}(Y),$$

in order to have $\mathrm{Var}(Z - Y)$ much less than $\mathrm{Var}(Z)$. This technique is particularly easy to implement when $\mathbf{E}(Y) = 0$. We easily construct, with the help of the Brownian motion, random variables with zero average. Indeed, if $(H(s), 0 \leq s \leq T)$ is an adapted process such that $\mathbf{E}\left(\int_0^T H(s)^2\,ds\right) < +\infty$, we have (see Section 5.1.3):

$$\mathbf{E}\left(\int_0^T H(s)\,dB(s)\right) = 0.$$

Theorem 5.4.1 proves that we can hope to annul in theory (and diminish in practice) the variance of the random variable Z with the help of a control variable that is a stochastic integral.

Theorem 5.4.1 *Let Z be a random variable such that $\mathbf{E}(Z^2) < +\infty$. Assume further that Z is measurable with respect to the algebra $\sigma(B(s), s \leq T)$. Then, there exists a process $(H(t), t \leq T)$ adapted to $\sigma(B(s), s \leq t)$, satisfying $\mathbf{E}\left(\int_0^T H(s)^2\,ds\right) < +\infty$ and such that:*

$$Z = \mathbf{E}(Z) + \int_0^T H(s)\,dB(s).$$

This theorem is a variant of the theorem called the 'predicted representation' theorem whose proof we find in Karatzas and Shreve (1988) or Revuz and Yor (1991).

Remark We note that Z must be measurable with respect to the *natural filtration* of the Brownian motion.

Theorem 5.4.1 allows us to confirm that, in principle, we can annul the variance of Z. However, the explicit calculation of $(H(s), s \leq T)$ is delicate (in fact, more complicated than the calculation of $\mathbf{E}(Z)$!). We find a formula for $H(s)$ using the Malliavin gradient and the conditional expectation in Newton (1994). This formula is, however, not easily exploitable (see Newton, 1994, for general approximation methods). These methods are often computationally intensive and we have often to resort to empirical techniques guided by the practical problem.

We can however show that, for certain parabolic problems whose representation is obtained by the Feynman–Kac theorem, the process $(H(t), t \leq T)$ can be put in the form $H(t) = v(t, X(t))$, v being a function of t and x. Theorem 5.4.2 clarifies this point.

Theorem 5.4.2 *Let b and σ be two Lipschitz functions, let $(X(t), t \geq 0)$ be the unique solution of*

$$dX(t) = b(X(t))\, dt + \sigma(X(t))\, dB(t), \quad X(0) = x.$$

We denote by A the infinitesimal generator of the diffusion:

$$Af(x) = \frac{1}{2} \sum_{i,j=1}^{n} a_{ij}(x) \frac{\partial^2 f}{\partial x_i \partial x_j}(x) + \sum_{j=1}^{n} b_j(x) \frac{\partial f}{\partial x_j}(x),$$

where $a_{ij}(x) = \sum_{k=1}^{p} \sigma_{ik}(x)\sigma_{jk}(x)$.

Assume that u is a function of class $C^{1,2}$ having bounded derivatives in x, which is the solution of the following partial differential equation:

$$\begin{cases} \left(\dfrac{\partial u}{\partial t} + Au \right)(t, x) = f(x) & \text{for } (t, x) \in [0, T] \times \mathbf{R}^n, \\ u(T, x) \qquad\qquad\ = g(x) & \text{for } x \in \mathbf{R}^n. \end{cases}$$

Then if $Z = g(X(T)) - \int_0^T f(X(s))\, ds$ and $Y = \int_0^T (\partial u/\partial x)(s, X(s))\sigma(s, X(s))\, dB(s)$, we have:

$$\mathbf{E}(Z) = Z - Y.$$

This means that the random variable Y provides a perfect control variable for Z.

Proof By applying the Itô formula to $u(t, X(t))$, we obtain:

$$du(t, X(t)) = \left(\frac{\partial u}{\partial t} + Au \right)(t, X(t))\, dt + \frac{\partial u}{\partial x}(t, X(t))\sigma(s, X(s))\, dB(t).$$

By integrating between 0 and T, and taking expected values of the two sides of the equation and since u is solution of the partial differential equation, we obtain:

$$u(0, x) = Z - Y = \mathbf{E}(Z).$$

$\square$

Remark When we want to calculate $\mathbf{E}(Z)$ to obtain the solution of a parabolic equation, Theorem 5.4.2 allows us to restrict the class in which we look for $H(t)$. Obviously, the explicit formula involving a partial derivative in x is not easily exploitable as it requires the resolution of a problem more complicated than the initial calculation of expectation.

In practice, we can proceed in the following way. If we want to calculate $u(0, x) = \mathbf{E}(Z)$ with $Z = g(X(T)) - \int_0^T f(X(s))\, ds$ and we know a rough approximation $\bar{u}$ of u it is natural to set as a control variable:

$$Y = \int_0^T \frac{\partial \bar{u}}{\partial x}(t, X(t))\sigma(s, X(s))\, dB(t).$$

Note that whatever the choice of $\bar{u}$, we obtain an unbiased estimate of $\mathbf{E}(Z)$ by setting $Z' = Z - Y$. If the choice of $\bar{u}$ is reasonable we can hope to improve significantly the variance of the estimator.

5.4.2 *Example of the use of control variables*

Example 1 Imagine that we want to calculate an option price in a Black–Scholes model whose volatility σ is random. This means that the price of the asset $(S(t), t \geq 0)$ is the solution of the stochastic differential equation:

$$dS(t) = S(t)\left(r\, dt + \sigma(t)\, dB(t)\right), \quad S(0) = x$$

and that $\sigma(t)$ is the solution of another stochastic differential equation, for example:

$$d\sigma(t) = b(\sigma(t))\, dt + c(\sigma(t))\, dB'(t), \quad \sigma(0) = \sigma$$

$(B(t), t \geq 0)$ and $(B'(t), t \geq 0)$ being two independent Brownian motions. We want to calculate:

$$\mathbf{E}\left(e^{-rT} f(S(T))\right).$$

If the expected variation $\sigma(t)$ is not too important, it is possible to use $e^{-rT} f(\bar{S}(T))$ as a control variable, with $\bar{S}(T)$ the solution of:

$$dS(t) = S(t)\left(r\, dt + \sigma\, dB(t)\right), \quad S(0) = x.$$

Example 2 We sometimes use the following method for calculations of financial products. Assume that the asset $(S(t), t \geq 0)$, which carries the optional product is the solution of a stochastic differential equation. Assume, further, that the price of the option that we want to calculate can be expressed in the form $C(t, S(t))$ (which is very common). Often, we can find a rough approximation of $C(t, x)$ by a function that we can calculate explicitly $\bar{C}(t, x)$. We then proceed in the following way, we simulate the process S at times $(t_k = kh, 0 \le k \le N)$ with the help of, for example, the Euler scheme $(\bar{S}(kh), 0 \le k \le N)$. We can then use as control variable:

$$Y = \sum_{k=1}^{N} \frac{\partial \bar{C}}{\partial x}(t_k, \bar{S}_{t_k})\left(\left(\bar{S}_{t_{k+1}} - \bar{S}_{t_k}\right) - \mathbf{E}\left(\bar{S}_{t_{k+1}} - \bar{S}_{t_k}\right)\right).$$

It is often easy to calculate $\mathbf{E}\left(\bar{S}_{t_{k+1}} - \bar{S}_{t_k}\right)$, which leads to an explicit control variable Y with zero expectation. If $\bar{C}$ is close to C, and if N is sufficiently large, we can expect a significant gain for the Monte-Carlo method.

Example 3 In this example, we want to calculate the average price of an option. We have seen, in Section 5.2.6, that this price is written in the form:

$$M = \mathbf{E}\left(e^{-rT}\left(\frac{1}{T}\int_0^T S(s)\,ds - K\right)_+\right),$$

where S is the Black–Scholes process:

$$S(t) = x\exp\left(\left(r - \frac{\sigma^2}{2}\right)t + \sigma B(t)\right).$$

When σ and r are not too large, we can hope that:

$$\frac{1}{T}\int_0^T S(s)\,ds \quad \text{'is not too far' from} \quad \exp\left(\frac{1}{T}\int_0^T \log(S(s))\,ds\right).$$

This heuristic argument allows us to think that the random variable:

$$Y = e^{-rT}\left(\exp(Z) - K\right)_+,$$

with $Z = (1/T)\int_0^T \log(S(s))\,ds$, can play the role of a control variable. Since the random variable Z is a Gaussian variable, we know how to calculate explicitly:

$$\mathbf{E}\left(e^{-rT}\left(\exp(Z) - K\right)_+\right).$$

This method, proposed by Kemma and Vorst (1990), is particularly efficient when σ is of the order of 0.5, r of the order of 0.1, and T of the order of 1 year. (These orders of magnitude are typical of those used for financial modelling.) For larger values of σ or for larger values of T the gain is less notable.

5.4.3   Importance function and Girsanov's theorem

Another method can be easily applied to diffusion problems, it is a variant of the technique of importance functions (see Section 1.4). In effect, we can use a classical result from diffusion theory, the Girsanov theorem, to identify interesting classes of importance functions.

For a diffusion, we can construct a method of preferential sampling in the following way. We are given a functional $\psi(x(s), 0 \leq s \leq T)$ of the path of a diffusion process. We then want to evaluate:

$$\mathbf{E}(Z),$$

where $Z = \psi(X(s), 0 \leq s \leq T)$ and $(X(s), s \geq 0)$ is a diffusion process constructed on a probability space $\mathbf{P}$. We shall define a new probability measure $\tilde{\mathbf{P}}$ starting from $\mathbf{P}$ by setting:

$$d\tilde{\mathbf{P}} = \theta d\mathbf{P},$$

θ being a strictly positive random variable with integral 1. When we calculate an expected value under the new probability $\tilde{\mathbf{P}}$, we denote this expectation by $\tilde{\mathbf{E}}$. We have, therefore, for every positive or integrable random variable Y:

$$\tilde{\mathbf{E}}(Y) = \mathbf{E}(\theta Y),$$

or again

$$\mathbf{E}(Z) = \tilde{\mathbf{E}}(\theta^{-1} Z).$$

To use this type of operator transformation in simulation, we must be able to simulate the product $\theta^{-1} Z$ under the new probability $\tilde{\mathbf{P}}$. In the case of diffusion processes, it is Girsanov's theorem that will allow us to do this.

Theorem 5.4.3 (Girsanov's theorem) *Let $(B(t), t \geq 0)$ be a Brownian motion with values in $\mathbf{R}^n$ with respect to a filtration $(\mathcal{F}_t, t \geq 0)$. Let $(h(t), t \leq T)$ be a process with values in $\mathbf{R}^n$ adapted to the filtration $\mathcal{F}_t$ and such that, almost surely:*

$$\int_0^T |h(s)|^2 \, ds < +\infty.$$

We set:

$$\mu_T = \exp\left(\int_0^T h(s) \, dB(s) - \frac{1}{2} \int_0^T |h(s)|^2 \, ds\right)$$

and

$$\tilde{B}(t) = B(t) - \int_0^T h(s) \, ds.$$

Then, if $\mathbf{E}(\mu_T) = 1$ and $\tilde{\mathbf{P}} = \mu_T \mathbf{P}$, the process $(\tilde{B}(t), 0 \leq t \leq T)$ is a Brownian motion under the probability $\tilde{\mathbf{P}}$.

We can find a proof of this result in Karatzas and Shreve (1988) or Revuz and Yor (1991).

Remark We can verify that $\mathbf{E}(\mu_T) = 1$ if $(h(t), t \geq 0)$ is a bounded process.

More generally, we can show that this result remains true if there exists a constant $c > 0$ such that:

$$\sup_{0 \leq t \leq T} \mathbf{E}\left(e^{c|h(t)|^2}\right) < +\infty.$$

Assume that $(X(t), t \geq 0)$ is the unique solution of

$$dX(t) = b(X(t))\, dt + \sigma(X(t))\, dB(t), \quad X(0) = x,$$

with b and σ Lipschitz functions and $(B(t), t \geq 0)$ a Brownian motion. If we are given a process $(h(t), t \geq 0)$ such that $\mathbf{E}(\mu_T) = 1$, then X is also a solution of

$$dX(t) = (b(X(t))\, dt - \sigma(X(t))h(t))\, dt + \sigma(X(t))\, d\tilde{B}(t), \quad X(0) = x,$$

with $(\tilde{B}(t), 0 \leq t \leq T)$, which is a Brownian motion under the probability $\tilde{\mathbf{P}}$. It is therefore possible to simulate the process X under the new probability $\tilde{\mathbf{P}}$. This is particularly easy when $h(t)$ is of the form $v(t, X(t))$, v being a function of t and of x, since X satisfies, then, a classical stochastic differential equation. In this case, since $\mathbf{E}(Z) = \tilde{\mathbf{E}}\left(\mu_T^{-1} Z\right)$, we can use the preceding transformation to construct a method of variance reduction.

Proposition 5.4.4 shows that we can, under very broad hypotheses, hope to annul the variance of the new random variable $\mu_T^{-1} Z$ and give some ideas about how to diminish it in practice.

Proposition 5.4.4 *Let Z be a random variable of the form $Z = \psi(X(s), 0 \leq s \leq T)$ such that $\mathbf{E}\left(Z^2\right) < +\infty$ and $\mathbf{P}\left(Z \geq \epsilon\right) = 1$, for an $\epsilon > 0$. We define $h(t)$ by*

$$h(t) = -\frac{H(t)}{\mathbf{E}\left(Z|\mathcal{F}_t\right)},$$

where $(H(t), 0 \leq t \leq T)$ an adapted process satisfying:

$$Z = \mathbf{E}\left(Z\right) + \int_0^T H(s)\, dB(s).$$

Set:

$$\mu_T = \exp\left(-\int_0^T h(s)\, dB(s) - \frac{1}{2}\int_0^T |h(s)|^2\, ds\right),$$

then $\mathbf{E}(\mu_T) = 1$ and we have:

$$\mathbf{E}\left(Z\right) = \tilde{\mathbf{E}}\left(\mu_T^{-1} Z\right) \quad and \quad \tilde{\mathbf{P}}\left(\mu_T^{-1} Z = \mathbf{E}\left(Z\right)\right) = 1.$$

Remark From the point of view of Monte-Carlo methods, this means that, if we know how to simulate $\mu_T^{-1} Z$ under $\tilde{\mathbf{P}}$, we therefore have an estimator of zero variance.

Proof Set:

$$\phi(t) = \frac{\mathbf{E}\left(Z | \mathcal{F}_t\right)}{\mathbf{E}\left(Z\right)}.$$

We have:

$$\phi(t) = 1 + \int_0^t \frac{H(s)}{\mathbf{E}\left(Z\right)}\, dB(s) = 1 - \int_0^t \phi(s) h(s)\, dB(s).$$

We deduce that almost surely under $\mathbf{P}$, and also under $\tilde{\mathbf{P}}$:

$$\phi(T) = \exp\left(-\int_0^T h(s)\, dB(s) - \frac{1}{2}\int_0^T |h(s)|^2\, ds\right) = \mu_T.$$

As $\phi(T) = Z/\mathbf{E}\left(Z\right)$ we deduce the result.   □

Remark Obviously, the effective calculation of $h(t)$ cited in Theorem 5.4.3 is very difficult. As in the control variable method case, we use intuitive methods to find acceptable candidates to reduce the variance. We shall find some approximation techniques in Newton (1994).

Remark In certain cases, as in the case of control variables, we can prove that the process $h(t)$ is of the form $v(t, X(t))$. We consider, for example, the important case where we want to evaluate:

$$\mathbf{E}\left(f(X(T))\right),$$

f being a positive bounded function. We have seen that if u is a function of class $C^{1,2}$, having bounded derivatives in x is the solution of the following partial differential equation:

$$\begin{cases} u(T, x) & = f(x) \quad \text{for } x \in \mathbf{R}^n, \\ \left(\dfrac{\partial u}{\partial t} + Au\right)(t, x) = 0 \quad \text{for } (t, x) \in [0, T] \times \mathbf{R}^n, \end{cases}$$

where A is the differential operator associated with the diffusion X, starting from the point x at the initial time, we have:

$$u(t, X(t)) = u(0, x) + \int_0^T \frac{\partial u}{\partial x}(s, X(s)) \sigma(X(s))\, dB(s).$$

From the martingale property of the stochastic integral we deduce that $(u(t, X(t)), 0 \le t \le T)$ is a martingale. This allows us to confirm that:

$$\mathbf{E}\left(f(X(T)) | \mathcal{F}_t\right) = \mathbf{E}\left(u(T, X(T)) | \mathcal{F}_t\right) = u(t, X(t)).$$

We therefore have, in particular,

$$f(X(T)) = \mathbf{E}\left(f(X(T))\right) + \int_0^T \frac{\partial u}{\partial x}(s, X(s))\sigma(X(s))\, dB(s).$$

As further $\mathbf{E}\left(f(X(T))|\mathcal{F}_t\right) = u(t, X(t))$ we can make explicit (as a function of u) the process $h(t)$ which annuls the variance:

$$h(t) = -\frac{(\partial u/\partial x)(s, X(s))\sigma(X(s))}{u(t, X(t))}.$$

In particular, $h(t)$ is of the form $v(t, X(t))$.

In this case, if we know a very rough approximation $\bar{u}(t, x)$ of $u(t, x)$, a natural way to reduce the variance is to substitute the function $\bar{u}$ for u in the preceding formula. We can consult, for this subject, Fourni *et al.* (1996), which shows how techniques of large variation make it possible to obtain such an approximation and its use to reduce the variance.

5.5   Bibliographic comments

For a general introduction to random processes and, in particular, to diffusions see Bouleau (1988). For more details on Brownian motion and diffusion processes one can consult Karatzas and Shreve (1988) or Revuz and Yor (1991). The links between diffusion processes and partial differential equations are treated in depth in Friedman (1975), Bensoussan and Lions, (1978), Durrett (1984), and Dautray (1989). The reader interested in applications in finance can consult Section 5.8 of Karatzas and Shreve (1988) and Lamberton and Lapeyre, (1991). On the discretizations of stochastic differential equations we refer to Pardoux and Talay (1985), Talay (1995) and to pages 148–196 of Graham *et al.* (1996). To date, Kloeden and Platen, (1992) is the only book dedicated to this subject. We shall find in Bouleau and Talay (1992) and Graham *et al.* (1996) some examples of the representation of nonlinear partial differential equations using diffusion processes and associated Monte-Carlo methods.

REFERENCES

Alcouffe, R., Dautray, R., Forster, A., Ledanois, G., and Mercier, B. (ed.) (1985). *Monte-Carlo Methods and Applications in Neutronics*, Lectures Notes in Physics. Springer, Berlin.

Alouges, F. (1993). Implementation sur ipsc 860 d'une mthode de Monte-Carlo pour la rsolution de l'quation de Boltzmann. Technical Report 2724, Note CEA.

Arkeryd, L. (1981). Intermolecular forces of infinite range and the Boltzmann equation. *Arch. Rational Mech. Anal.* **77**, 11–21.

Babovsky, H. (1986). On a simulation scheme for the Boltzmann equation. *Math. Meth. Appl. Sci.* **8**, 223–233.

Bensoussan, A. and Lions, J.L. (1978). *Application des inquations variationnelles en contrle stochastique.* Dunod, Paris.

Bensoussan, A. and Lions, J.L. (1982). *Contrle impulsionnel et inquations quasivariationnelles.* Dunod, Paris.

Binder, K. (1984). *Application of the Monte-Carlo Methods in Statistical Physics.* Springer Verlag, Berlin.

Birdsall, C. (1991). Particle in cell charged particles simulations. *IEEE Trans. Plasma Sc.*, **19**, p 65.

Bird, G.A. (1963). Direct simulation Monte-Carlo method. *Phys. of Fluids* **6**, 1518.

Bird, G.A. (1976). *Molecular Gas Dynamics.* Clarendon Press, Oxford.

Booth, T.E. (1985). A sample problem for variance reduction in m.c.n.p. Technical Report 10363, Los Alamos Report.

Borgnakke, C. and Larsen, P.S. (1975). Statistical collision model for Monte Carlo simulations. *J. Comput. Phys.* **18**, 405–420.

Bouleau, N. (1986). *Probabilits de l'Ingnieur.* Hermann, Paris.

Bouleau, N. (1988). *Processus Stochastiques et Applications.* Hermann, Paris.

Bouleau, N. and Lepingle, D. (1993). *Numerical Methods for Stochastic Process.* John Wiley and Son, Inc, New York.

Bouleau, N. and Talay, D. (ed.) (1992). *Probabilits numriques.* INRIA.

Bourgat, J.F., Desvilettes, L., Tallec, P. Le, and Perthame, B. (1994). Microreversible collisions for polyatomic gases and Boltzmann's theorem. *Eur. J. Mech. B/Fluids.*

Bourgat, J.F., Tallec, P. Le, Tidriri, D., and Qiu, Y. (1992). Numerical coupling of nonconservative or kinetic models with Navier–Stokes. In *Domain Decomposition Methods for P.D.E.* (ed. D. Keyes and T. Chan). SIAM, Philadelphia.

Bratley, P., Fox, B.L., and Schrage, E.L. (1987). *A Guide to Simulation* (2nd edn). Springer Verlag, New York.

Breiman, L. (1968). *Probability*. Addison-Wesley, Reading, MA.

Brmaud, P. (1981). *Point Processes and Queues*. Springer, Berlin.

Caflisch, R. E. (1980). The fluid dynamic limit of the non linear Boltzmann equation. *Commun. Pure Appl. Math.* **33**, 651–666.

Cercignani, C. (1988). The Boltzmann equation and its applications. *Appl. Math.* **67**.

Cercignani, C., Illner, R., and Pulverenti, M. (1994). *The Mathematical Theory of Dilute Gases*. Number 106 in Applied Mathematical Sciences. Springer-Verlag, New York.

Cinlar, E. (1975). *Introduction to Stochastic Processes*. Prentice Hall, Englewood Cliffs, NJ.

Clark, J.M.C. and Cameron, R.J. (1980). The maximum rate of convergence of discrete approximation for stochastic differential equations. In *Lecture Notes in Control and Information Sciences, Stochastic Differential Systems* (ed. B.Grigelionis), Volume 25, Berlin. Springer Verlag, Berlin.

Cochran, W.G. (1977). *Sampling Techniques*. John Wiley and Sons, New York.

Dautray, R. (ed.) (1989). *Mthodes Probabilistes pour les quations de la physique*. Collection CEA. Eyrolles, Paris.

Dautray, R. and Lions, J.L. (1994). *Analyse mathmatique et calcul numrique*. Masson, Paris. Tome 9 : numrique, transport.

Desvilettes, L. and Peralta, R. (1995). A vectorizable simulation method for the Boltzmann equation. *Math. Modelling and Numerical Anal.*.

Devroye, L. (1986). *Non Uniform Random Variate Generation*. Springer Verlag, New York.

Durrett, R. (1984). *Brownian Motion and Martingales in Analysis*. Wadsworth, Belmont, CA.

Ethier, S. and Kurtz, T. (1986). *Markov Processes*. Wiley, New York.

Faure, O. (1992a). Numerical pathwise approximation of stochastic differential equation. *Appl. Stochastic Models Data Anal.*

Faure, O. (1992b). *Simulation du mouvement brownien et des diffusions*. Ph. D. thesis, Thse de Doctorat de l'Ecole Nationale des Ponts et Chausses.

Fourni, E., Lasry, J.M., and Touzi, N. (1996). Large deviations, small noise expansion and variance reduction. *Preprint*.

Friedman, A. (1975). *Stochastic Differential Equations and Application*, Volume Volumes 1 and 2. Academic Press, New York.

Giorla, J. and Sentis, R. (1987). A random walk method for solving radiative transfer equations. *J. Comp. Phys.* **70**, 145–165.

Graham, C., Kurtz, T., Mlard, S., Protter, P., Pulvirenti, M., and Talay, D. (1996). *Probabilistic Models for Nonlinear PDE's and Numerical*

Applications. Number 1627 in Lecture Notes in Mathematics. Springer-Verlag. CIME Summer School, D. Talay and L. Tubaro (eds).

Graham, C. and Mlard, S. (1995). Convergence rate on path space for stochastic particle approximations to the Boltzmann equation. Technical report, CMAP—Ecole Polytechnique.

Gropengiesser, F., Neunzert, H., and Struckmeier, J. (1990). Computational methods for the the Boltzmann equations. In *The State of Art in Applied and Industrial Maths* (ed. R. Spigler). Kluver Academic, Dordrecht.

Grunbaum, F.A. (1971). Propagation of chaos for the Boltzmann equation. *Arch. Rational Mech. Anal.* **42**, 323–345.

Hammersley, J.M. and Handscomb, D.C. (1964). *Monte Carlo Methods.* Chapman and Hall, London.

Hockney, R.W. and Eastwood, J.R. (1981). *Computer Simulation using Particles.* Mc Graw Hill.

Illner, R. and Neunzert, H. (1987). On the simulation methods for the Boltzmann equation. *Transport Theory Stat. Phys.* **16**, 141–154.

Ivanov, M.S. and Rogasinsky, S.V. (1988). Analysis of numerical techniques of the direct Monte Carlo simulation method. *Sov. J. Numer. Anal. Math. Modelling* **3**, 453–465.

Kac, M. (1956). Foundations of kinetic theory. proc. In *Proceedings of the Third Berkeley Symposium on Math. Statistics,* Volume III. Univ Calif.

Kalos, M.H. and Whitlock, P.A. (1986). *Monte Carlo Methods, volume I: Basics.* John Wiley and Sons.

Karatzas, I. and Shreve, S.E. 1988. *Brownian Motion and Stochastic Calculus.* Springer-Verlag, New York.

Karlin, S. and Taylor, H. (1981). *A Second Course in Stochastic Processes.* Academic Press, New York.

Kemma, A.G.Z and Vorst, A.C.F. (1990, March). A pricing method for options based on average asset values. *J. Banking Finan.* 113–129.

Kloeden, P.E. and Platen, E. (1992). *Numerical Solution of Stochastic Differential Equations.* Springer Verlag, Berlin.

Knuth, D.E. (1981). *The Art of Computer Programming, Seminumerical Algorithms,* Volume 2. Addison-Wesley, Reading, MA.

Kuipers, L. and Neiderreiter, H. (1974). *Uniform Distribution of Sequences.* Wiley, New York.

Kushner, H.J. (1977). *Probability Methods for Approximations in Stochastic Control and for Elliptic Equations.* Academic Press, New York.

Kushner, H.J. (1990). *Weak Convergence Methods and Singularly Perturbed Stochastic Control and Filtering Problems.* Birkhuser, Basel.

Lamberton, D. and Lapeyre, B. (1991). *Une Introduction au calcul Stochastique Applique la Finance.* Collection Mathmatiques et Applications. Ellipse.

L'Ecuyer, P. (1990, Octobre). Random numbers for simulation. *Commu. ACM* **33**(10).

Metropolis, N. and Ulam, S.M. (1949).   The Monte Carlo method. *J. Amer. Statist. Assoc.* **44**, 335–349.

Mischler, S. and Wennberg, B. (1996). On the homogeneous Boltzmann equation. *Prepublication du laboratoire d'Analyse Numrique de Paris 6.*

Morgan, B.J.T. (1984).   *Elements of Simulation.*   Chapman and Hall, London.

Morgenstern, D. (1955). Analytical studies related to the Maxwell–Boltzmann equation. *J. Rat. Mech. Anal.* **4**, 533–555.

Morokoff, W.J. and Caflish, R.E. (1994, November). Quasi random sequences and their discrepancies. *SIAM J. Sci. Comput.* **15**(6), 1251–1279.

Nanbu, K. (1980). Direct simulation schemes derived from the Boltzmann equation. *J. Phys. Japan* **49**, 2042.

Neiderreiter, H. (1992). *Random Number Generation and Quasi Monte Carlo Methods.* Society for Industrial and Applied Mathematics.

Neveu, J. and Pardoux, E. (1992). Modles de diffusion. Cours de l'Ecole Polytechnique.

Newton, N.J. (1994). Variance reduction for simulated diffusions. *SIAM J. Appl. Math.* **54**(6), 1780–1805.

N'Kaoua, T. and Sentis, R. (1993). A new time discretization for the radiative tranfer equations. *SIAM J. Numer. Anal.* **30**, 733–748.

Novikov, A.A. (1972). On an identity for stochastic integrals. *Theory Probab. Appl.* **17**, 717–720.

Papanicolaou, G.C. (1975). Asymptotic analysis of transport processes. *Bull. Amer. Math. Soc.* **81**, 330–392.

Pardoux, E. and Talay, D. (1985). Approximation and simulation of solutions of stochastic differential equation. *Acta Applicandae Math.* **3**, 23–47.

Pardoux, E. (1993). Evolutions alatoires et quations de transport. Cours de l'Ecole Polytechnique.

Perna, R. Di and Lions, P.L. (1989). On the Cauchy problem for Boltzmann equations. *Ann. Math.* **130**, 321–366.

Pinsky, M. (1991). *Lectures on Random Evolution.* World Scientific, Singapore.

Press, W.H., Teukolsky, S.A., Flannery, B.P., and Vetterling, W.T. (1992). *Numerical Recepies.* Cambridge University Press, Cambridge.

Pulvirenti, M., Wagner, W., and Rossi, M.B. Zavelani (1994). Convergence of particle schemes for the Boltzmann equation. *Eur. J. Mech. B/Fluids*, **13**.

Raviart, P.A. (1985). An analysis of particle methods. In *Numerical Methods in Fluid Dynamics* (ed. F. Brezzi), Volume 1127 of *Lectures Notes in Mathematics.* Springer, Berlin.

Revuz, D. and Yor, M. (1991). *Continuous Martingales and Brownian Motion.* Springer Verlag, Berlin.

Ringoissen, F. and Sentis, R. (1991). On the diffusion approximation of a transport process without time scaling. *Asymptotic Anal.* 5, 145–159.

Ripley, B.D. (1987). *Stochastic Simulation.* John Wiley and Sons, New York.

Rogers, L.C.G. and Shi, Z. (1995). The value of an asian option. *J. Appl. Probab.* 32(4), 1077–1088.

Rogers, L.C.G. and Williams, D. (1994). *Diffusions, Markov Processes and Martingales, Volume 1, Foundations, Second Edition.* John Wiley and Sons, New York.

Rubinstein, R. Y. (1981). *Simulation and the Monte Carlo Method.* John Wiley and Sons, New York.

Sedgewick, R. (1987). *Algorithms.* Addison-Wesley, Reading, MA.

Sentis, R. and Dellacherie, S. Sur un traitement de collisions ractives ou nuclaires par des oprateurs de type Boltzmann. Note CEA.

Spanier, J. and Gelbard, E. (1969). *Monte Carlo principles and Neutron Transport Problems.* Series in Computer Science and Information Processing. Addison-Wesley, Reading, MA.

Spanier, J. and Maze, E. (1994). Quasi-random methods for estimating integrals using relatively small samples. *Siam Rev.* 36(1), 18–44.

Sznitman, A.S. (1984). Equations de type de Boltzmann spatialement homognes. *Z. Wahrscheinlichkeitstheorie v. Gebiete* 66, 559–592.

Talay, D. (1986). Discrtisation d'une quation diffrentielle stochastique et calcul approch d'esprances de fonctionnelles de la solution. *Math. Modelling Numer. Anal.* 20, 141–179.

Talay, D. (1995). Simulation and numerical analysis of stochastic differential systems: a review. In *Probabilistic Methods in Applied Physics* (ed. P. Kre and W. Wedig), Volume 451 of *Lecture Notes in Physics*, Chapter 3, pp. 54–96. Springer-Verlag, Berlin.

Wagner, W. (1988). Monte Carlo evaluation of functionals of stochastic differential equations-variance reduction and numerical examples. *Stochastic Anal. Appl.* 6, 447–468.

Wagner, W. (1989). Unbiased Monte-Carlo estimators for fonctionals of weak solutions of stochastic differential equations. *Stochastic and Stochastics Reports* 28, 1–20.

Wagner, W. (1992). A convergence proof for Bird's Monte Carlo method for the Boltzmann equation. *J. Statist. Phys.* 66, 1011–1044.

INDEX

Printed in the United States
by Bookmasters

Printed in the United States
By Bookmasters